Workbook

Division

A Direct Instruction Program

Siegfried Engelmann • Doug Carnine

Columbus, OH

The McGraw·Hill Companies

SRAonline.com

Division Preskill Test

A Subtraction

```
   60        486        307        402        452
 - 32      - 189      - 245      - 386      - 180
```

```
  356        500        454        478        375
- 186      - 364      - 327      - 296      - 278
```

B Multiplication

```
   35         54         46         54         34
 ×  8      ×   7      ×   6      ×   4      ×   3
```

```
   94         75         95         37         85
 ×  9      ×   6      ×   8      ×   7      ×   6
```

A Subtraction

```
  60        486        307        402        452
- 32      - 189      - 245      - 386      - 180


 356        500        454        478        375
- 186      - 364      - 327      - 296      - 278
```

B Multiplication

```
  35         54         46         54         34
×  8       ×  7       ×  6       ×  4       ×  3


  94         75         95         37         85
×  9       ×  6       ×  8       ×  7       ×  6
```

Division Placement Test

5)15   1)7   3)6   9)36   1)8

9)18   3)12   1)3   5)25   5)5

3)6   9)9   5)10   9)27   3)9

5)20   9)45   1)4   3)3   3)15

B

5)70   9)216   5)116   3)104   3)45

3 buses left Midville each day. 12 buses left
in all. How many days did buses leave
Midville?

☐ days

Every time Fred went jogging, he ran 5
blocks. He ran 10 blocks. How many times
did he go jogging?

☐ times

Part B continues on the next page.

Division Placement Test (continued)

Pete did 10 problems on each page. He did 5 pages. How many problems did he do?

[ ] problems

Betty typed 2 pages each hour. She typed 8 pages. How many hours did she type?

[ ] hours

There are 6 apples in each pile. There are 2 piles of apples. How many apples in all?

[ ] apples

C

$48\overline{)264}$ $27\overline{)162}$ $82\overline{)2354}$

$54\overline{)3267}$ $74\overline{)7934}$ $73\overline{)7142}$

Lesson 1

Bonus

1

A
$$2\overline{)6}$$ — quotient 3

B
$$5\overline{)10}$$ — quotient 2

C
$$2\overline{)12}$$ — quotient 6

2

A
$$5\overline{)15}$$

B
$$5\overline{)10}$$

C
$$5\overline{)20}$$

D
$$5\overline{)5}$$

3

A
$$9\overline{)45}$$

B
$$9\overline{)9}$$

C
$$9\overline{)27}$$

D
$$9\overline{)18}$$

4

A
$$4\overline{)24}$$ — quotient 6 $$4\overline{)24}$$ quotient 6 $$6\overline{)24}$$ quotient 4

B
$$3\overline{)15}$$ — quotient 5

C
$$4\overline{)12}$$ — quotient 3

5

A
$$5\overline{)25}$$

B
$$9\overline{)18}$$

C
$$9\overline{)27}$$

D
$$5\overline{)15}$$

Lesson 2

Bonus

1

A   5⟌25

B   5⟌10

C   5⟌20

D   5⟌15

2

A   9⟌36

B   9⟌18

C   9⟌45

D   9⟌27

3

A

B

C

4

A   5⟌2͞0 ⁴⟍

B   9⟌3͞6 ⁴⟍

C   9⟌1͞8 ²⟍

D   5⟌1͞0 ²⟍

2 •——— Lesson 2

Lesson 3

Bonus

1

A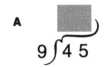
9 ⟌ 4 5

B
9 ⟌ 9

C
9 ⟌ 2 7

D
9 ⟌ 1 8

2

A
5 ⟌ 2 5

B
5 ⟌ 1 0

C
5 ⟌ 5

D
5 ⟌ 2 0

3

A B C

4

A
$\frac{2}{9 ⟌ 1\,8}$

B
$\frac{4}{9 ⟌ 3\,6}$

C
$\frac{3}{5 ⟌ 1\,5}$

D
$\frac{1}{5 ⟌ 5}$

5

A
5 ⟌ 5

B
5 ⟌ 1 0

C
5 ⟌ 1 5

D
5 ⟌ 2 0

E
5 ⟌ 2 5

Lesson 3 ■ **3**

1

A 5⟌5

B 5⟌10

C 5⟌15

D 5⟌20

2

A 1⟌4

B 1⟌10

C 1⟌7

D 1⟌5

3

A	**B**	**C**	**D**	**E**
2⟌6	5⟌10	2⟌12	6⟌18	7⟌14

| | | | | |

4

A 1⟌7

B 1⟌4

C 1⟌2

D 1⟌6

Lesson 5

Bonus

1

A 5)25

B 5)10

C 5)5

D 5)20

2

A **B** **C** **D**

3

A 1)7

B 1)4

C 1)9

D 1)6

4

A 8)16

B 2)10

C 4)12

D 9)9

E 2)8

||||||||||||||||

5

A 1)10

B 1)8

C 1)1

D 1)9

Lesson 6

Facts + Bonus = TOTAL

1

A $1\overline{)3}$

B $1\overline{)10}$

C $1\overline{)8}$

D $1\overline{)1}$

2

A   B   C   D

3

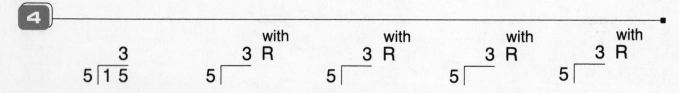

(0) 1 2 3 4 (5) 6 7 8 9 (10) 11 12 13 14 (15) 16 17 18 19 (20) 21 22 23 24 (25)

4

$$5\overline{)15}^{\;3}$$ $$5\overline{)}^{\;3\;R}\text{with}$$ $$5\overline{)}^{\;3\;R}\text{with}$$ $$5\overline{)}^{\;3\;R}\text{with}$$ $$5\overline{)}^{\;3\;R}\text{with}$$

5

A $9\overline{)27}$

B $9\overline{)36}$

C $9\overline{)18}$

D $9\overline{)45}$

6

A $5\overline{)25}$

B $5\overline{)15}$

C $5\overline{)5}$

D $5\overline{)10}$

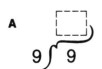

Facts + Bonus = TOTAL

1

A
9⟌9

B
9⟌18

C
9⟌27

D
9⟌36

2

⓪ 1 2 3 4 ⑤ 6 7 8 9 ⑩ 11 12 13 14 ⑮ 16 17 18 19 ⑳ 21 22 23 24 ㉕

3

2
5⟌10

2 with R
5⟌$$

2 with R
5⟌$$

2 with R
5⟌$$

2 with R
5⟌$$

4

A
9⟌18

B
9⟌9

C
9⟌27

D
9⟌36

5

A
1⟌7

B
1⟌10

C
1⟌4

D
1⟌5

Lesson 8

1

A 9⟌9

B 9⟌18

C 9⟌27

D 9⟌36

2

A 5⟌21

B 5⟌12

C 5⟌16

D 5⟌9

E 5⟌19

3

A 9⟌36

B 9⟌45

C 9⟌18

D 9⟌27

4

A 5⟌5

B 5⟌20

C 5⟌15

D 5⟌10

5

Write all the numbers 5 goes into 3 times **with** a remainder.

$\dfrac{3}{5⟌15}$ A $5⟌$ 3 R   B $5⟌$ 3 R   C $5⟌$ 3 R   D $5⟌$ 3 R

Problems + Bonus = TOTAL

1

A
9⟌27

B
9⟌9

C
9⟌18

D
9⟌45

2

A
5⟌11

B
5⟌22

C
5⟌7

D
5⟌13

E
5⟌17

3

A
```
     4
5⟌21
  -20
     1
```

B
5⟌23

C
5⟌17

D
5⟌14

E
5⟌18

F
5⟌12

4

A
$$\frac{4}{5\overline{)2\ 2}}$$
$-\underline{}$

B
$$\frac{2}{5\overline{)1\ 1}}$$
$-\underline{}$

C
$$\frac{4}{5\overline{)2\ 3}}$$
$-\underline{}$

D
$$\frac{3}{5\overline{)1\ 9}}$$
$-\underline{}$

E
$$\frac{1}{5\overline{)8}}$$
$-\underline{}$

F
$$\frac{3}{5\overline{)1\ 6}}$$
$-\underline{}$

G
$$\frac{2}{5\overline{)1\ 2}}$$
$-\underline{}$

H
$$\frac{4}{5\overline{)2\ 4}}$$
$-\underline{}$

5

Write all the numbers 5 goes into 4 times **with** a remainder.

$$\frac{4}{5\overline{)2\ 0}}$$

A
$$\frac{4\ \text{R}}{5\overline{)}}$$

B
$$\frac{4\ \text{R}}{5\overline{)}}$$

C
$$\frac{4\ \text{R}}{5\overline{)}}$$

D
$$\frac{4\ \text{R}}{5\overline{)}}$$

Write all the numbers 5 goes into 2 times **with** a remainder.

$$\frac{2}{5\overline{)1\ 0}}$$

E
$$\frac{2\ \text{R}}{5\overline{)}}$$

F
$$\frac{2\ \text{R}}{5\overline{)}}$$

G
$$\frac{2\ \text{R}}{5\overline{)}}$$

H
$$\frac{2\ \text{R}}{5\overline{)}}$$

Lesson 10

1

A
9⌐18

B
9⌐45

C
9⌐36

D
9⌐9

2

A   B   C   D

3

A
5⌐16

B
5⌐24

C
5⌐13

D
5⌐18

E
5⌐8

4

A
5⌐17

B
5⌐14

C
5⌐27

D
5⌐8

E
5⌐22

5

A
$$5\overline{|9}^{\,1}$$
—

B
$$5\overline{|1\,3}^{\,2}$$
—

C
$$5\overline{|2\,4}^{\,4}$$
—

D
$$5\overline{|6}^{\,1}$$
—

E
$$5\overline{|2\,7}^{\,5}$$
—

F
$$5\overline{|2\,3}^{\,4}$$
—

G
$$5\overline{|1\,9}^{\,3}$$
—

H
$$5\overline{|1\,2}^{\,2}$$
—

6

$$5\overline{|1\,5}$$ $$9\overline{|1\,8}$$ $$9\overline{|4\,5}$$ $$1\overline{|1\,0}$$ $$5\overline{|2\,5}$$ $$9\overline{|3\,6}$$ $$9\overline{|9}$$

$$9\overline{|3\,6}$$ $$1\overline{|7}$$ $$5\overline{|2\,0}$$ $$5\overline{|1\,5}$$ $$1\overline{|2}$$ $$5\overline{|5}$$ $$5\overline{|2\,5}$$

7

Write all the numbers 5 goes into 1 time **with** a remainder.

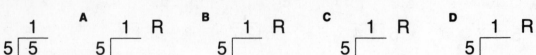

$$5\overline{|5}^{\,1}$$ A $$5\overline{|}^{\,1\,R}$$ B $$5\overline{|}^{\,1\,R}$$ C $$5\overline{|}^{\,1\,R}$$ D $$5\overline{|}^{\,1\,R}$$

Write all the numbers 5 goes into 4 times **with** a remainder.

$$5\overline{|2\,0}^{\,4}$$ E $$5\overline{|}^{\,4\,R}$$ F $$5\overline{|}^{\,4\,R}$$ G $$5\overline{|}^{\,4\,R}$$ H $$5\overline{|}^{\,4\,R}$$

1

⓪ 1 2 3 4 5 6 7 8 ⑨ 10 11 12 13 14 15 16 17 ⑱ 19 20 21 22 23 24 25 26 ㉗

2

$$9\overline{)1\,8}^{\,2}$$ $$9\overline{)}^{\,2\ R}$$ $$9\overline{)}^{\,2\ R}$$ $$9\overline{)}^{\,2\ R}$$ $$9\overline{)}^{\,2\ R}$$

$$9\overline{)}^{\,2\ R}$$ $$9\overline{)}^{\,2\ R}$$ $$9\overline{)}^{\,2\ R}$$ $$9\overline{)}^{\,2\ R}$$

3

A $$5\overline{)1\,7}$$

B $$5\overline{)8}$$

C $$5\overline{)2\,3}$$

D $$5\overline{)9}$$

E $$5\overline{)1\,2}$$

F $$5\overline{)1\,9}$$

4

A $$5\overline{)1\,8}$$ **B** $$9\overline{)1\,9}$$ **C** $$5\overline{)1\,7}$$ **D** $$9\overline{)6}$$ **E** $$9\overline{)3\,8}$$ **F** $$5\overline{)6}$$

5

A
$$\begin{array}{r} 2 \\ 9\overline{)2\,2} \\ -1\,8 \end{array}$$

B
$$\begin{array}{r} 3 \\ 9\overline{)3\,3} \\ -2\,7 \end{array}$$

C
$$\begin{array}{r} 4 \\ 9\overline{)4\,4} \\ -3\,6 \end{array}$$

6

A The big number in a times problem is 10. One small number is 5.

B One small number in a times problem is 10. The other small number is 5.

C One small number in a times problem is 9. The other small number is 4.

D The big number in a times problem is 36. One small number is 9.

E The big number in a times problem is 20. One small number is 5.

7

$5\overline{)1\,0}$ $9\overline{)1\,8}$ $9\overline{)9}$ $1\overline{)8}$ $9\overline{)3\,6}$ $5\overline{)2\,5}$ $5\overline{)1\,5}$

$9\overline{)2\,7}$ $5\overline{)5}$ $1\overline{)4}$ $5\overline{)1\,0}$ $1\overline{)6}$ $5\overline{)2\,0}$ $9\overline{)1\,8}$

8

Finish working these problems and show the remainders.

A
$9\overline{)2\,6}$ with quotient 2

B
$5\overline{)1\,6}$ with quotient 3

C
$5\overline{)2\,9}$ with quotient 5

D
$9\overline{)1\,2}$ with quotient 1

E
$5\overline{)1\,2}$ with quotient 2

F
$5\overline{)8}$ with quotient 1

G
$9\overline{)3\,8}$ with quotient 4

H
$5\overline{)2\,9}$ with quotient 5

I
$5\overline{)1\,8}$ with quotient 3

J
$9\overline{)4\,7}$ with quotient 5

Lesson 12

1

⓪ 1 2 3 4 5 6 7 8 ⑨ 10 11 12 13 14 15 16 17 ⑱ 19 20 21 22 23 24
25 26 ㉗ 28 29 30 31 32 33 34 35 ㊱ 37 38 39 40 41 42 43 44 ㊺ 46 47 . . .

2

$$9\overline{)27} \;3$$

$$9\overline{)}\;3\text{ R}$$

$$9\overline{)}\;3\text{ R}$$

$$9\overline{)}\;3\text{ R}$$

$$9\overline{)}\;3\text{ R}$$

$$9\overline{)}\;3\text{ R}$$

$$9\overline{)}\;3\text{ R}$$

$$9\overline{)}\;3\text{ R}$$

$$9\overline{)}\;3\text{ R}$$

3

A
$5\overline{)21}$

B
$5\overline{)24}$

C
$5\overline{)13}$

D
$5\overline{)18}$

E
$5\overline{)6}$

F
$5\overline{)17}$

4

A $9\overline{)38}$ B $5\overline{)14}$ C $5\overline{)23}$ D $9\overline{)17}$ E $9\overline{)29}$

5

A
$$9\overline{)52}\;\;5$$
$$-\underline{45}$$

B
$$9\overline{)23}\;\;2$$
$$-\underline{18}$$

C
$$9\overline{)40}\;\;4$$
$$-\underline{36}$$

6

A One small number in a times problem is 5. The other small number is 3.

B The big number in a times problem is 45. One small number is 9.

7

A The big number in a times problem is 20. One small number is 5. What's the third number?

B One small number in a times problem is 3. The other small number is 4. What's the third number?

C One small number in a times problem is 1. The other small number is 8. What's the third number?

D The big number in a times problem is 18. One small number is 9. What's the third number?

8

$5\overline{)5}$ $9\overline{)27}$ $9\overline{)36}$ $5\overline{)15}$ $9\overline{)9}$ $9\overline{)45}$ $1\overline{)6}$

$1\overline{)1}$ $9\overline{)18}$ $5\overline{)10}$ $1\overline{)9}$ $5\overline{)25}$ $9\overline{)36}$ $5\overline{)20}$

$9\overline{)27}$ $5\overline{)15}$ $9\overline{)18}$ $5\overline{)20}$ $9\overline{)45}$ $5\overline{)25}$ $1\overline{)8}$

9

Finish working these problems and show the remainders.

A
$$5\overline{)8} \quad \begin{array}{r} 1 \\ \hline \end{array}$$

B
$$9\overline{)38} \quad \begin{array}{r} 4 \\ \hline \end{array}$$

C
$$5\overline{)29} \quad \begin{array}{r} 5 \\ \hline \end{array}$$

D
$$5\overline{)18} \quad \begin{array}{r} 3 \\ \hline \end{array}$$

E
$$9\overline{)47} \quad \begin{array}{r} 5 \\ \hline \end{array}$$

1

A
9 ⌐1 9⌐

B
9 ⌐3 0⌐

C
9 ⌐1 2⌐

D
9 ⌐2 4⌐

E
9 ⌐4 1⌐

2

A
```
    4
5 ⌐1 8
- 2 0
```

B
```
    5
9 ⌐4 1
- 4 5
```

C
```
    3
5 ⌐1 0
- 1 5
```

D
```
    3
9 ⌐2 9
- 2 7
```

E
```
    4
9 ⌐3 0
- 3 6
```

F
```
    2
5 ⌐1 4
- 1 0
```

3

A One small number in a times problem is 6. The other small number is 3. What's the third number?

B One small number in a times problem is 2. The other small number is 8. What's the third number?

C The big number in a times problem is 27. One small number is 9. What's the third number?

D The big number in a times problem is 15. One small number is 5. What's the third number?

E The big number in a times problem is 36. One small number is 9. What's the third number?

F One small number in a times problem is 7. The other small number is 5. What's the third number?

4

$5\overline{)20}$ $9\overline{)27}$ $1\overline{)4}$ $5\overline{)5}$ $9\overline{)18}$ $5\overline{)10}$ $5\overline{)15}$

$9\overline{)9}$ $9\overline{)36}$ $5\overline{)10}$ $9\overline{)45}$ $5\overline{)15}$ $9\overline{)18}$ $9\overline{)36}$

5

Finish working these problems and show the remainders.

A
$$4\overline{)23} \quad 5$$
$$-\underline{}$$

B
$$9\overline{)49} \quad 5$$
$$-\underline{}$$

C
$$3\overline{)28} \quad 9$$
$$-\underline{}$$

6

Write all the numbers 9 goes into 3 times **with** a remainder.

$$9\overline{)27} \quad 3$$

A $9\overline{)}\ 3\ R$ B $9\overline{)}\ 3\ R$ C $9\overline{)}\ 3\ R$ D $9\overline{)}\ 3\ R$

E $9\overline{)}\ 3\ R$ F $9\overline{)}\ 3\ R$ G $9\overline{)}\ 3\ R$ H $9\overline{)}\ 3\ R$

7

Write the facts with no remainders.

A $5\overline{)6}$ $\overline{)}$

B $5\overline{)17}$ $\overline{)}$

C $5\overline{)12}$ $\overline{)}$

D $5\overline{)23}$ $\overline{)}$

E $5\overline{)14}$ $\overline{)}$

F $5\overline{)19}$ $\overline{)}$

8

Write the answer. Then multiply and subtract to find the remainder.

A	B	C	D	E
5$\overline{)17}$	9$\overline{)31}$	9$\overline{)20}$	5$\overline{)8}$	9$\overline{)14}$

F	G	H	I	J
9$\overline{)21}$	5$\overline{)9}$	5$\overline{)16}$	9$\overline{)40}$	9$\overline{)19}$

9

A 5$\overline{)5}$

B 5$\overline{)20}$

C 5$\overline{)15}$

D 5$\overline{)10}$

E 9$\overline{)18}$

F 9$\overline{)45}$

G 9$\overline{)36}$

H 9$\overline{)9}$

Lesson 14

1

A
$3\overline{)15}$

B
$3\overline{)9}$

C
$3\overline{)6}$

D
$3\overline{)12}$

2

A
$3\overline{)3}$

B
$3\overline{)6}$

C
$3\overline{)9}$

D
$3\overline{)12}$

3

A
$9\overline{)21}$

B
$9\overline{)12}$

C
$9\overline{)39}$

D
$9\overline{)32}$

E
$9\overline{)11}$

4

A
$\begin{array}{r} 3 \\ 9\overline{)19} \\ -27 \end{array}$

B
$\begin{array}{r} 5 \\ 9\overline{)43} \\ -45 \end{array}$

C
$\begin{array}{r} 4 \\ 5\overline{)23} \\ -20 \end{array}$

D
$\begin{array}{r} 2 \\ 9\overline{)16} \\ -18 \end{array}$

E
$\begin{array}{r} 5 \\ 5\overline{)26} \\ -25 \end{array}$

F
$\begin{array}{r} 3 \\ 5\overline{)11} \\ -15 \end{array}$

5

A
There are 4 blocks in each pile. There are 16 blocks in all.

B
Every time Sam goes to the store, he buys 6 carrots. He has 42 carrots in all.

C
There are 9 towels in each box. There are 27 towels in all.

D
Mary Jo builds 4 benches each day that she works. She builds 20 benches in all.

E
Every time Mrs. Whitehead goes jogging, she runs 5 kilometers. She runs 35 kilometers in all.

6

$3\overline{)12}$ $5\overline{)25}$ $3\overline{)15}$ $9\overline{)45}$ $3\overline{)9}$ $9\overline{)18}$ $3\overline{)12}$

$3\overline{)9}$ $3\overline{)15}$ $9\overline{)36}$ $5\overline{)20}$ $3\overline{)12}$ $9\overline{)27}$ $3\overline{)9}$

7

Write all the numbers 9 goes into 4 times **with** a remainder.

$\begin{array}{r} 4 \\ 9\overline{)36} \end{array}$

A $9\overline{)}\ 4\ R$

B $9\overline{)}\ 4\ R$

C $9\overline{)}\ 4\ R$

D $9\overline{)}\ 4\ R$

E $9\overline{)}\ 4\ R$

F $9\overline{)}\ 4\ R$

G $9\overline{)}\ 4\ R$

H $9\overline{)}\ 4\ R$

8

Write the facts with no remainders.

A
5⟌1 4

B
5⟌1 6

C
5⟌1 9

D
5⟌2 4

E
5⟌1 1

F
5⟌8

9

Write the fact for each problem.

A One small number in a times problem is 2. The other small number is 4. What's the third number?

B The big number in a times problem is 18. One small number is 9. What's the third number?

C The big number in a times problem is 20. One small number is 5. What's the third number?

D One small number in a times problem is 2. The other small number is 5. What's the third number?

10

Write the answer. Then multiply and subtract to find the remainder.

A
9⟌4 0

B
5⟌1 6

C
9⟌2 1

D
5⟌9

E
9⟌1 9

Facts + Problems + Bonus = TOTAL

1

A 3⟌6

B 3⟌15

C 3⟌3

D 3⟌12

2

A 3⟌15

B 3⟌6

C 3⟌9

D 3⟌12

3

A 9⟌29

B 9⟌31

C 9⟌13

D 9⟌38

E 9⟌20

4

A
```
   4
9⟌21
 −36
```

B
```
   5
5⟌23
 −25
```

C
```
   3
9⟌22
 −27
```

D
```
   4
5⟌14
 −20
```

E
```
   5
9⟌40
 −45
```

F
```
   2
9⟌13
 −18
```

5

A

There are 5 paper cups in each pile. There are 25 paper cups in all.

B

Every time Ned goes to the store, he buys 3 oranges. He has 12 oranges in all.

6

A

Every time Carlos walked to school, he walked 5 blocks. Carlos walked 200 blocks in all. How many times did Carlos walk to school?

B

Every time Carlos walked to school, he walked 5 blocks. Carlos walked to school 200 times. How many blocks did Carlos walk in all?

C

Every time Pam's team scored, they got 7 points. Pam's team scored 75 times. How many points did they get in all?

D

Every time Yuki used pens, she used 4 pens. Yuki used 12 pens in all. How many times did Yuki use pens?

7

$3\overline{)12}$ $5\overline{)15}$ $3\overline{)9}$ $3\overline{)15}$ $1\overline{)7}$ $3\overline{)12}$ $5\overline{)25}$

$5\overline{)5}$ $3\overline{)15}$ $3\overline{)12}$ $1\overline{)10}$ $9\overline{)9}$ $3\overline{)15}$ $3\overline{)9}$

8

Write the fact for each problem.

A The big number in a times problem is 20. One small number is 5. What's the third number?

B One small number in a times problem is 5. The other small number is 4. What's the third number?

C The big number in a times problem is 45. The other number is 9. What's the third number?

9

Write all the numbers 9 goes into 5 times **with** a remainder.

$$\begin{array}{r} 5 \\ 9\overline{\smash{)}4\,5} \end{array}$$

A $9\overline{\smash{)}}\ 5\ R$ **B** $9\overline{\smash{)}}\ 5\ R$ **C** $9\overline{\smash{)}}\ 5\ R$ **D** $9\overline{\smash{)}}\ 5\ R$

E $9\overline{\smash{)}}\ 5\ R$ **F** $9\overline{\smash{)}}\ 5\ R$ **G** $9\overline{\smash{)}}\ 5\ R$ **H** $9\overline{\smash{)}}\ 5\ R$

10

Write the facts with no remainders.

A $5\overline{\smash{)}1\,8}$ **B** $5\overline{\smash{)}1\,1}$ **C** $5\overline{\smash{)}7}$

D $5\overline{\smash{)}2\,1}$ **E** $5\overline{\smash{)}1\,2}$ **F** $5\overline{\smash{)}1\,7}$

11

Write the answer. Then multiply and subtract to find the remainder.

A $5\overline{\smash{)}1\,8}$ **B** $9\overline{\smash{)}2\,0}$ **C** $5\overline{\smash{)}6}$ **D** $9\overline{\smash{)}4\,1}$ **E** $9\overline{\smash{)}2\,2}$

Lesson 16

1

A
$$3\overline{)6}$$

B
$$3\overline{)15}$$

C
$$3\overline{)3}$$

D
$$3\overline{)9}$$

2

A
$$9\overline{)20}$$

B
$$9\overline{)29}$$

C
$$9\overline{)22}$$

D
$$9\overline{)14}$$

E
$$9\overline{)39}$$

F
$$9\overline{)15}$$

3

A
$$\begin{array}{r} 3 \\ 5\overline{)13} \\ -15 \end{array}$$

B
$$\begin{array}{r} 3 \\ 9\overline{)16} \\ -27 \end{array}$$

C
$$\begin{array}{r} 3 \\ 5\overline{)11} \\ -15 \end{array}$$

D
$$\begin{array}{r} 4 \\ 5\overline{)18} \\ -20 \end{array}$$

E
$$\begin{array}{r} 3 \\ 9\overline{)23} \\ -27 \end{array}$$

F
$$\begin{array}{r} 2 \\ 9\overline{)17} \\ -18 \end{array}$$

A Every day Mattie read 3 books. Mattie read 18 books in all. How many days did Mattie read books?

B Every time Jack fed the dogs, he used 7 cans of food. He fed the dogs 28 times. How many cans of food did Jack use in all?

C Every week Ron ate 5 bananas. Ron ate bananas for 50 weeks. How many bananas did Ron eat in all?

D Every time Ann made a dress, she used 3 meters of cloth. Ann used 45 meters of cloth in all. How many dresses did Ann make?

A
Don made 40 sandwiches each week. He made sandwiches for 13 weeks.

B
Each day Mary planted 6 trees. Mary has planted trees for 13 days.

C
Kit loaded 6 trucks every day. She loaded trucks for 13 days.

D
Don washed 10 tables each hour. He washed tables for 5 hours.

E
Every afternoon that Peg looked for worms, she found 12. Peg looked for worms for 5 afternoons.

6

$9\overline{)18}$ $3\overline{)6}$ $3\overline{)12}$ $5\overline{)10}$ $3\overline{)9}$ $3\overline{)6}$ $5\overline{)25}$

7

Write the fact for each problem.

A The big number in a times problem is 20. One small number is 5. What's the third number?

B One small number in a times problem is 5. The other small number is 5. What's the third number?

C The big number in a times problem is 30. One small number is 5. What's the third number?

D The big number in a times problem is 18. One small number is 9. What's the third number?

8

A
$5\overline{)7}$

B
$9\overline{)40}$

C
$9\overline{)19}$

D
$5\overline{)28}$

E
$9\overline{)15}$

F
$5\overline{)17}$

G
$9\overline{)30}$

H
$5\overline{)6}$

I
$9\overline{)46}$

J
$9\overline{)20}$

K
$9\overline{)14}$

L
$5\overline{)24}$

M
$9\overline{)38}$

N
$5\overline{)27}$

O
$5\overline{)11}$

1

A 5 ⟌ 3 5

B 5 ⟌ 5 0

C 5 ⟌ 3 0

D 5 ⟌ 4 5

2

A 5 ⟌ 3 0

B 5 ⟌ 3 5

C 5 ⟌ 4 0

D 5 ⟌ 4 5

3

A 3 ⟌ 1 5

B 3 ⟌ 9

C 3 ⟌ 6

D 3 ⟌ 1 2

4

A 9 ⟌ 1 2

B 9 ⟌ 1 9

C 9 ⟌ 2 5

D 9 ⟌ 3 9

E 9 ⟌ 2 8

F 9 ⟌ 1 5

5

A
```
      3
 4 | 1 7
   - 1 2
       5
```

B
```
        5
 9 | 5 4
   - 4 5
        9
```

C
```
        6
 3 | 2 0
   - 1 8
        2
```

D
```
        9
 5 | 5 0
   - 4 5
        5
```

6

A
Each day Maria drank 3 glasses of milk. Maria drank 15 glasses of milk.

B
Each day Jessie bought 9 apples. Jessie bought 27 apples.

C
Fred washed 9 floors every week. Fred washed floors for 18 weeks.

D
Lucy used 5 pens in every class. Lucy used pens in 10 classes.

E
Every month Joe watered his plants 5 times. He watered his plants 10 times.

7

3 | 9   5 | 3 0   3 | 6   5 | 4 0   5 | 3 5   3 | 1 2   3 | 1 5

5 | 4 0   3 | 6   5 | 3 5   3 | 1 2   5 | 3 0   3 | 6   5 | 3 5

5 | 3 0   3 | 1 5   3 | 9   5 | 4 0

8

Finish the problems. If you can't subtract, make the answer smaller. Then copy the problem and the new answer in the space next to the problem.

A
```
      3
 9 | 2 2
   - 2 7
```

B
```
      4
 5 | 2 4
   - 2 0
```

C
```
      5
 5 | 1 3
   - 2 5
```

D
```
      4
 9 | 2 4
   - 3 6
```

Lesson 17 (continued)

9

Write the fact for each problem.

A The big number in a times problem is 9. One small number is 3. What's the third number?

B One small number in a times problem is 9. The other small number is 2. What's the third number?

C One small number in a times problem is 5. The other small number is 4. What's the third number?

D The big number in a times problem is 20. One small number is 5. What's the third number?

10

A	B	C	D	E
3⟌1 1	9⟌3 0	5⟌2 1	3⟌5	9⟌3 4

F	G	H	I	J
3⟌1 4	5⟌2 0	3⟌8	9⟌2 0	3⟌1 2

K	L	M	N	O
9⟌3 8	3⟌1 6	5⟌1 1	3⟌1 3	9⟌2 9

Lesson 18

Facts + Problems + Bonus = TOTAL

1

A 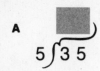 5⟌3 5

B 5⟌4 0

C 5⟌5 0

D 5⟌4 5

2

A 5⟌3 0

B 5⟌3 5

C 5⟌4 0

D 5⟌4 5

3

A   B   C   D

4

A
```
    5
8⟌4 8
 -4 0
    8
```

B
```
    6
4⟌2 9
 -2 4
    5
```

C
```
    4
3⟌1 6
 -1 2
    4
```

D
```
    6
7⟌4 9
 -4 2
    7
```

E
```
    3
9⟌3 2
 -2 7
    5
```

F
```
    6
5⟌3 0
 -3 0
    0
```

5

A

A dog chewed 6 bones every day. It chewed bones for 90 days. How many bones did the dog chew?

B

George cleaned 9 erasers each week. He cleaned 45 erasers. How many weeks did George clean erasers?

C

Marcy washed 3 cars every hour. She washed cars for 60 hours. How many cars did Marcy wash?

$9\overline{)27}$ $5\overline{)35}$ $3\overline{)12}$ $5\overline{)30}$ $5\overline{)40}$ $9\overline{)36}$ $5\overline{)40}$

$5\overline{)30}$ $3\overline{)6}$ $5\overline{)35}$ $3\overline{)9}$ $5\overline{)40}$ $5\overline{)30}$ $5\overline{)35}$

Write the facts with no remainders.

A $9\overline{)40}$

B $9\overline{)47}$

C $9\overline{)20}$

D $9\overline{)16}$

E $9\overline{)6}$

F $9\overline{)43}$

8

Finish the problems. If you can't subtract, make the answer smaller. Then copy the problem and the new answer in the space next to the problem.

A
$$5\overline{)12} \quad \begin{array}{r} 2 \\ \hline \end{array}$$
$$-10$$

B
$$9\overline{)17} \quad \begin{array}{r} 2 \\ \hline \end{array}$$
$$-18$$

C
$$5\overline{)4} \quad \begin{array}{r} 1 \\ \hline \end{array}$$
$$-5$$

D
$$9\overline{)46} \quad \begin{array}{r} 5 \\ \hline \end{array}$$
$$-45$$

9

A $9\overline{)20}$ **B** $5\overline{)17}$ **C** $3\overline{)8}$ **D** $3\overline{)10}$ **E** $9\overline{)31}$

Facts + Problems + Bonus = TOTAL

1

A

5⟌40

B

5⟌35

C

5⟌50

D

5⟌30

2

A
```
   4
4⟌2 1
- 1 6
   5
```

B
```
   4
7⟌3 2
- 2 8
   4
```

C
```
   7
6⟌5 0
- 4 2
   8
```

D
```
   5
3⟌1 9
- 1 5
   4
```

3

A

Shu fixed 4 tires every day. He fixed tires for 32 days. How many tires did Shu fix?

B

Mr. Silbert wrote 2 stories every week. He wrote 12 stories. How many weeks did he write stories?

C

Carmen painted 5 cars every week. She painted 10 cars. How many weeks did Carmen paint?

4

A

Cora put 7 books in each pile. She used 21 books. How many piles did Cora make?

B

Larry baked 5 cakes each week. He baked cakes for 75 weeks. How many cakes did Larry bake?

C

Ms. Hightower ate 5 oranges each week. She ate oranges for 10 weeks. How many oranges did Ms. Hightower eat?

D

The Big Steel Company made 2 sheets of steel every hour. They made sheets of steel for 40 hours. How many sheets of steel did they make?

E

Mike picked 3 flowers every afternoon. He ended up with 15 flowers. How many afternoons did Mike pick flowers?

5

$5\overline{)35}$ $5\overline{)45}$ $5\overline{)20}$ $5\overline{)40}$ $5\overline{)30}$ $5\overline{)15}$ $5\overline{)50}$

$3\overline{)6}$ $9\overline{)18}$ $5\overline{)45}$ $9\overline{)36}$ $5\overline{)50}$ $5\overline{)25}$ $5\overline{)45}$

$5\overline{)30}$ $3\overline{)15}$ $5\overline{)40}$ $5\overline{)50}$ $5\overline{)35}$

6

Write the facts with no remainders.

A $9\overline{)23}$

B $9\overline{)17}$

C $9\overline{)42}$

D $9\overline{)12}$

E $9\overline{)37}$

F $9\overline{)20}$

G $3\overline{)5}$

H $3\overline{)10}$

I $3\overline{)13}$

J $3\overline{)14}$

K $3\overline{)17}$

L $3\overline{)8}$

7

Write the fact for each problem.

A

The big number in a times problem is 9. One small number is 3. What is the third number?

B

The big number in a times problem is 10. One small number is 5. What is the third number?

C

One small number in a times problem is 10. The other small number is 5. What is the third number?

D

One small number in a times problem is 6. The other small number is 4. What is the third number?

E

The big number in a times problem is 15. One small number is 3. What is the third number?

F

One small number in a times problem is 5. The other small number is 3. What is the third number?

8

A	B	C	D	E
3$\overline{)11}$	5$\overline{)18}$	3$\overline{)14}$	3$\overline{)10}$	9$\overline{)31}$

Lesson 20

Test + Facts + Problems + Bonus = TOTAL

1

A 5 ⟌ 4 0

B 5 ⟌ 5 0

C 5 ⟌ 4 5

D 5 ⟌ 3 5

2

A
```
      7
8 ⟌ 6 6
  − 5 6
    1 0
```

B
```
      6
4 ⟌ 3 1
  − 2 4
      7
```

C
```
      8
7 ⟌ 6 3
  − 5 6
      7
```

3

A 8 ⟌ 0

B 9 ⟌ 0

C 4 ⟌ 0

D 6 ⟌ 0

E 3 ⟌ 0

4

A Kathy cut 3 lawns every week. She cut 12 lawns. How many weeks did Kathy work?

  _ _ _ _ _ _ _ _ _ _ _ _ _ _ _ _

B Joy put 5 hooks on each shelf. She put hooks on 90 shelves. How many hooks did Joy use?

  _ _ _ _ _ _ _ _ _ _ _ _ _ _ _ _

C Stan lost 3 golf balls every time he played golf. He lost 15 golf balls. How many times did Stan play golf?

  _ _ _ _ _ _ _ _ _ _ _ _ _ _ _ _

D Lamar won 9 prizes every time he read a book. He read 18 books. How many prizes did Lamar win?

  _ _ _ _ _ _ _ _ _ _ _ _ _ _ _ _

5

A	B	C	D	E
3⟌14	9⟌30	9⟌32	5⟌9	9⟌25

6

5⟌50	9⟌27	5⟌30	5⟌45	3⟌12	5⟌40	5⟌45
3⟌9	5⟌50	9⟌45	5⟌35	5⟌20	5⟌50	5⟌15
3⟌15	5⟌45	5⟌35	5⟌30	5⟌40	5⟌10	9⟌36

7

Write the facts with no remainders.

A
9⟌25

B
9⟌3

C
9⟌14

D
9⟌41

E
9⟌16

F
9⟌21

G
3⟌7

H
3⟌16

I
3⟌1

J
3⟌11

K
3⟌4

L
3⟌13

1

A	B	C	D	E
7 $\overline{\smash{)}0}$	2 $\overline{\smash{)}0}$	9 $\overline{\smash{)}0}$	5 $\overline{\smash{)}0}$	3 $\overline{\smash{)}0}$

2

A   B   C   D

3

A	B	C	D
5 $\overline{\smash{)}638}$	5 $\overline{\smash{)}4627}$	5 $\overline{\smash{)}532}$	9 $\overline{\smash{)}467}$

4

A A student used 3 papers every lesson. She did 50 lessons. How many papers did the student use?

B A seal balanced 5 balls in every show. It balanced 40 balls in shows. How many times was the seal in a show?

C Judy worked 3 hours every day. She worked 60 days. How many hours did Judy work?

D Jane's store was open 5 hours every week. Her store was open for 30 hours. How many weeks was Jane's store open?

E Kim cut her hair 3 times every year. She cut her hair 90 times. How many years did Kim cut her hair?

5

| 5⟌40 | 3⟌9 | 5⟌15 | 3⟌3 | 9⟌36 | 9⟌45 | 5⟌35 |

| 5⟌25 | 6⟌0 | 3⟌12 | 5⟌50 | 9⟌27 | 3⟌15 | 1⟌10 |

| 5⟌5 | 1⟌8 | 5⟌45 | 5⟌0 | 9⟌9 | 1⟌3 | 3⟌6 |

| 3⟌0 | 9⟌36 | 3⟌12 | 5⟌20 | 3⟌9 | 5⟌30 | 9⟌0 |

6

Write the facts with no remainders.

A 3⟌7

B 3⟌5

C 3⟌10

D 3⟌8

E 3⟌16

F 3⟌13

7

A 5⟌33   B 9⟌41   C 3⟌14   D 9⟌25   E 5⟌6   F 3⟌10

G 9⟌21   H 5⟌11   I 9⟌38   J 3⟌7   K 5⟌44   L 9⟌16

1

A 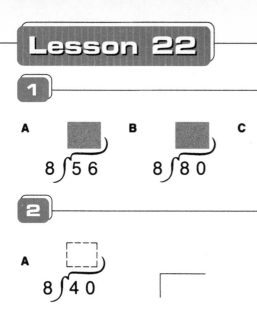 8 ⟌ 5 6   B 8 ⟌ 8 0   C 8 ⟌ 7 2   D 8 ⟌ 4 8   E 8 ⟌ 6 4   F 8 ⟌ 4 0

2

A 8 ⟌ 4 0

B 8 ⟌ 4 8

C 8 ⟌ 5 6

D 8 ⟌ 6 4

E 8 ⟌ 7 2

F 8 ⟌ 8 0

3

A 5 ⟌ 4 7 3   B 5 ⟌ 5 6 2 1   C 5 ⟌ 6 8 0   D 3 ⟌ 1 6 2

4

A Hank painted 3 houses every week. He painted 15 houses. How many weeks did Hank paint houses?

B Sally watches 5 trains each day. She watched 30 trains. How many days did Sally watch trains?

Part 4 continues on the next page.

c Rico took 8 piano lessons every month.
   He took them for 7 months. How many
   lessons did Rico take?

☐ _____

d Peg did 3 cartwheels for each cartoon
   show she watched on television. She did
   12 cartwheels. How many cartoons did
   Peg watch?

☐ _____

e A plane carried 3 people on each trip.
   It made 6 trips. How many people rode
   in the plane?

☐ _____

f Tony lost 9 pencils every year. He lost
   pencils for 3 years. How many pencils
   did Tony lose?

☐ _____

5

$9\overline{)36}$ $5\overline{)15}$ $3\overline{)3}$ $5\overline{)35}$ $9\overline{)27}$ $3\overline{)9}$ $5\overline{)50}$

$7\overline{)0}$ $5\overline{)30}$ $5\overline{)25}$ $3\overline{)15}$ $2\overline{)0}$ $9\overline{)45}$ $5\overline{)5}$

$5\overline{)40}$ $5\overline{)20}$ $3\overline{)6}$ $9\overline{)18}$ $5\overline{)10}$ $5\overline{)40}$ $3\overline{)12}$

6

$8\overline{)56}$ $8\overline{)40}$ $5\overline{)35}$ $3\overline{)6}$ $8\overline{)48}$ $5\overline{)40}$ $5\overline{)30}$

$5\overline{)50}$ $8\overline{)56}$ $5\overline{)45}$ $8\overline{)40}$ $3\overline{)15}$ $8\overline{)48}$ $8\overline{)40}$

7

Write the facts with no remainders.

A 5)4 7

B 5)3 6

C 5)4 3

D 5)4 2

E 5)3 4

F 5)4 6

G 5)3 9

H 5)3 1

I 5)4 4

8

A 3)1 1

B 9)3 6

C 9)3 1

D 5)5

E 9)2 0

F 3)4

G 5)1 3

H 3)6

I 9)4 7

J 3)1 7

K 5)1 8

L 3)8

M 9)3 1

N 3)1 4

O 5)2 7

Facts + Problems + Bonus = TOTAL

1

A 8⟌48 **B** 8⟌40 **C** 8⟌56 **D** 8⟌80 **E** 8⟌72 **F** 8⟌64

2

A 8⟌40

B 8⟌48

C 8⟌56

D 8⟌64

E 8⟌72

F 8⟌80

3

A 3⟌134 **B** 5⟌76 **C** 3⟌673 **D** 9⟌942 **E** 5⟌214 **F** 9⟌37

4

A
```
      4 2
 9 ⟌ 3 7 9
   - 3 6 ↓
      1 9
     1 8
```

B
```
      1
 5 ⟌ 7 5
   - 5
      2
```

C
```
      4
 3 ⟌ 1 2 4
   - 1 2
      0
```

5

A Each flower has 5 petals. There are 20 petals on the flowers. How many flowers are there?

☐ _____

B Each motorcycle driver owns 6 helmets. There are 12 motorcycle drivers. How many helmets do they own?

☐ _____

C There are 5 desks in each row. There are 15 rows. How many desks are there?

☐ _____

D Tam ate 3 meals every day. She ate 9 meals. How many days did Tam eat meals?

☐ _____

E Nora was in 9 plays every month. She was in plays for 10 months. How many plays was Nora in?

☐ _____

F The president shook hands 9 times every minute. She shook hands 36 times. How many minutes did the president shake hands?

☐ _____

G Rita read 5 comic books every hour. She read comics for 5 hours. How many comic books did Rita read?

☐ _____

H Every time Lou went shopping, he went to 6 stores. He went shopping 9 times. How many stores did Lou go to?

☐ _____

Part 5 continues on the next page.

1 Rosa told 3 jokes every time she went to a party. She told 9 jokes at parties. How many times did she go to a party?

☐ _____

6

5)30   9)27   3)12   1)7   5)20   5)35   5)45

4)0   3)6   5)10   10)0   5)35   3)15   9)36

1)1   5)25   9)18   8)0   3)9   5)15   1)2

5)40   9)45   5)50   5)45   9)36   5)30   8)0

7

Write the facts with no remainders.

A 5)38 ⌐‾‾‾

B 5)46 ⌐‾‾‾

C 5)33 ⌐‾‾‾

D 5)43 ⌐‾‾‾

E 5)31 ⌐‾‾‾

F 5)42 ⌐‾‾‾

G 5)39 ⌐‾‾‾

H 5)17 ⌐‾‾‾

I 5)48 ⌐‾‾‾

1

A 8⟌72

B 8⟌64

C 8⟌80

D 8⟌48

E 8⟌40

F 8⟌56

2

A 9⟌318   B 5⟌482   C 3⟌102   D 9⟌940   E 3⟌172   F 5⟌391

3

A
```
      9
5 ⟌ 4 8 7
  − 4 5
      3
```

B
```
      5
5 ⟌ 2 9 8
  − 2 5
      4
```

C
```
      3
9 ⟌ 2 7 9
  − 2 7
      0
```

D
```
      2
3 ⟌ 7 4
  − 6
      1
```

E
```
      1
9 ⟌ 9 9
  − 9
      0
```

4

A Linda put together 9 bikes every day. She put together 27 bikes. How many days did Linda put bikes together?

☐ _____

B There are 5 people on every team. There are 40 people. How many teams are there?

☐ _____

C There are 2 socks in every drawer. There are 4 drawers of socks. How many socks are there?

☐ _____

D The quarterback threw 3 passes every game. He threw 9 passes during games. How many games did he throw passes in?

☐ _____

E Jerry saw 8 movies every week. He saw 72 movies. How many weeks did Jerry go to the movies?

☐ _____

F There were 8 different bands in each parade. There were 4 parades. How many bands were in the parades?

☐ _____

G Each TV set has 5 knobs. There are 35 knobs on TV sets. How many TV sets are there?

☐ _____

H Mark sees 9 plants in every window. He sees 18 plants in windows. How many windows are there?

☐ _____

5

5⟌15	3⟌6	9⟌18	5⟌45	1⟌10	3⟌15	5⟌25
1⟌0	5⟌40	3⟌9	1⟌5	5⟌10	3⟌3	9⟌45
5⟌50	6⟌0	5⟌20	9⟌9	3⟌12	5⟌35	9⟌27
9⟌36	5⟌30	5⟌5	3⟌6	9⟌45	3⟌0	1⟌8

6

8⟌48	8⟌80	8⟌64	8⟌56	5⟌30	8⟌72	8⟌80
8⟌64	8⟌40	3⟌15	8⟌72	5⟌50	8⟌56	8⟌72
3⟌6	5⟌45	3⟌12	8⟌80	8⟌40	8⟌64	8⟌48

7

Write the facts with no remainders.

A 5⟌32

B 5⟌41

C 5⟌38

D 5⟌48

E 5⟌43

F 5⟌36

Facts + Problems + Bonus = TOTAL

1

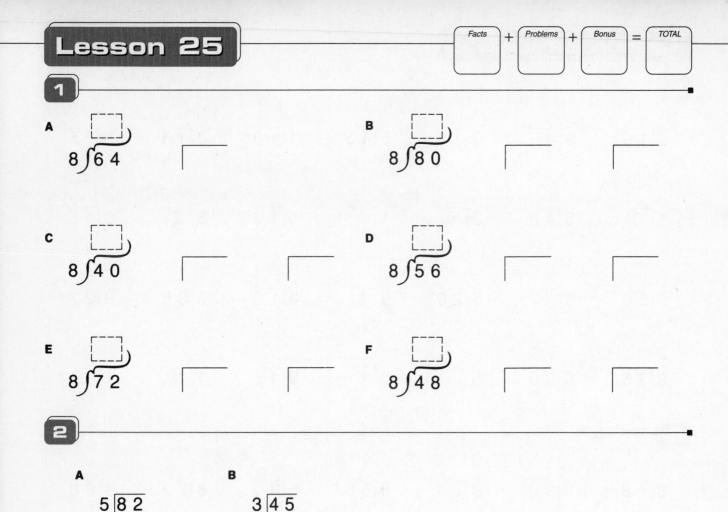

A 8)‾6 4

B 8)‾8 0

C 8)‾4 0

D 8)‾5 6

E 8)‾7 2

F 8)‾4 8

2

A 5)‾8 2

B 3)‾4 5

3

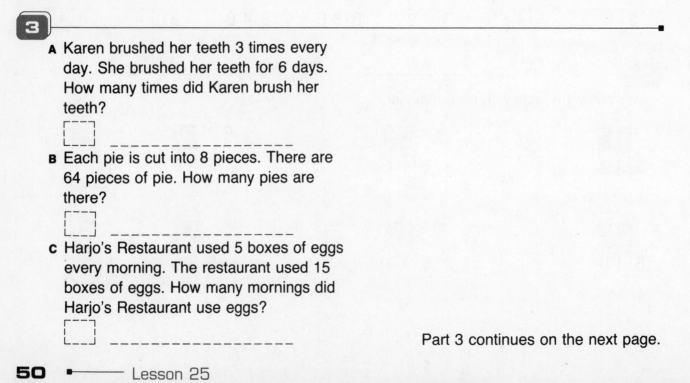

A Karen brushed her teeth 3 times every day. She brushed her teeth for 6 days. How many times did Karen brush her teeth?

B Each pie is cut into 8 pieces. There are 64 pieces of pie. How many pies are there?

C Harjo's Restaurant used 5 boxes of eggs every morning. The restaurant used 15 boxes of eggs. How many mornings did Harjo's Restaurant use eggs?

Part 3 continues on the next page.

D There were 3 chairs at each table. There are 9 tables. How many chairs are there?

☐ _____

E A horse ate 6 pails of oats each day. It ate oats for 12 days. How many pails of oats did the horse eat?

☐ _____

F Ken drove 8 kilometers every time he went to work. He drove 48 kilometers to work. How many times did Ken drive to work?

☐ _____

G Kevin cooked 3 hot dogs for each person at the party. He cooked 12 hot dogs. How many people were at the party?

☐ _____

4

5$\overline{)35}$	9$\overline{)18}$	3$\overline{)0}$	5$\overline{)15}$	9$\overline{)36}$	3$\overline{)9}$	3$\overline{)0}$
5$\overline{)25}$	5$\overline{)0}$	5$\overline{)40}$	3$\overline{)12}$	9$\overline{)0}$	1$\overline{)8}$	5$\overline{)20}$
3$\overline{)6}$	5$\overline{)50}$	9$\overline{)27}$	1$\overline{)6}$	5$\overline{)10}$	9$\overline{)0}$	5$\overline{)45}$
3$\overline{)15}$	5$\overline{)30}$	9$\overline{)45}$	5$\overline{)50}$	9$\overline{)27}$	3$\overline{)12}$	5$\overline{)0}$

5

8)64   3)9   8)40   8)72   8)56   8)80   8)48

8)72   5)35   8)64   5)50   8)56   8)80   5)45

5)30   3)12   8)80   3)6   8)72   8)40   8)64

6

A	B	C	D	E
9)35	3)14	9)40	5)17	3)11

7

Work only the first part of these problems. Underline the first part. Then find the answer and the remainder for the first part.

A	B	C	D	E
3)524	5)135	3)245	5)823	9)342

1

A   B   C   D

2

(50)   51   52   53   54   55   56   57   58   59   (60)

A 58 rounds to _____ tens.   **B** 53 rounds to _____ tens.

C 56 rounds to _____ tens.   **D** 57 rounds to _____ tens.

E 51 rounds to _____ tens.   **F** 54 rounds to _____ tens.

3

A   B   C   D   E   F

9⟌138   3⟌75   5⟌119   3⟌162   5⟌258   3⟌83

4

A Every classroom has 9 windows. There are 36 windows. How many classrooms are there?

B Every day Jackie loaded 5 trucks. She loaded trucks for 8 days. How many trucks did Jackie load?

C Joan used 8 pieces of cheese each time she made pizza. She used 64 pieces of cheese. How many pizzas did Joan make?

Part 4 continues on the next page.

D Each notebook has 32 pieces of paper.
There are 8 notebooks. How many pieces
of paper are there in all?

[ ] _____

5

5⟌35   3⟌12   8⟌48   9⟌27   5⟌45   8⟌72   5⟌25

8⟌56   8⟌40   9⟌0   5⟌5   3⟌6   5⟌50   9⟌36

5⟌50   5⟌45   3⟌9   8⟌64   8⟌48   5⟌30   8⟌72

9⟌9   5⟌40   8⟌80   9⟌45   5⟌20   5⟌15   5⟌0

8⟌80   5⟌10   8⟌64   5⟌40   8⟌56   3⟌0   8⟌40

3⟌15   5⟌30   3⟌3   9⟌18   5⟌35

6

Work only the first part of these problems. Find the first remainder and
then stop.

A   B   C   D   E
9⟌53   5⟌278   3⟌78   9⟌402   9⟌726

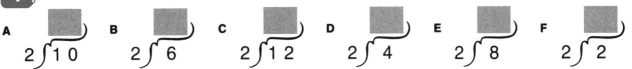
1

A 2⟌10   B 2⟌6   C 2⟌12   D 2⟌4   E 2⟌8   F 2⟌2

2

A 2⟌2

B 2⟌4

C 2⟌6

D 2⟌8

E 2⟌10

F 2⟌12

3

㉚ 31 32 33 34 <u>35</u> 36 37 38 39 ㊵

A 36 rounds to _____ tens.

B 33 rounds to _____ tens.

C 34 rounds to _____ tens.

D 32 rounds to _____ tens.

E 37 rounds to _____ tens.

F 39 rounds to _____ tens.

4

A   B   C

5⟌2 8 4 6   9⟌4 7 3 0   3⟌7 3 0 5

Part 4 continues on the next page.

D   E   F

9|3790   9|1231   5|2173

5

A 3 girls can fit into every tent. 12 girls are
going camping with tents. How many
tents will they need?

B Each branch has 8 leaves. There are 72
leaves. How many branches are there?

C Each goal is worth 3 points. Fast Freddy
scored 12 points. How many goals did
he make?

D Tim went to the grocery store 5 times
every month. He went to the grocery store
for 10 months. How many times did Tim
go to the grocery store?

E Paula worked 5 days each week. She
worked for 25 days. How many weeks did
Paula work?

6

5)25	8)48	5)30	1)10	9)0	5)20	8)80
9)36	5)0	8)72	8)40	3)12	5)10	8)56
5)25	8)64	3)6	7)0	5)45	8)56	9)27
5)30	8)72	9)0	5)5	1)4	9)18	5)35
3)15	5)40	8)64	9)45	3)9	8)48	5)20
3)3	5)50	5)15	3)0	9)9		

7

2)6	8)56	2)8	2)4	8)40	2)6	8)72
2)4	8)64	8)80	2)8	8)48	2)4	2)6

8

Write the facts with no remainders.

A ☐
8)45 ☐

B ☐
8)75 ☐

C ☐
8)86 ☐

D ☐
8)67 ☐

E ☐
8)70 ☐

F ☐
8)52 ☐

Lesson 28

1

A 2⟌8   B 2⟌1 2   C 2⟌6   D 2⟌1 0   E 2⟌4   F 2⟌2

2

A 2⟌2   B 2⟌4

C 2⟌6   D 2⟌8

E 2⟌1 0   F 2⟌1 2

3

A 36 rounds to _____ tens.   B 47 rounds to _____ tens.

C 93 rounds to _____ tens.   D 86 rounds to _____ tens.

E 11 rounds to _____ tens.   F 23 rounds to _____ tens.

G 78 rounds to _____ tens.   H 61 rounds to _____ tens.

I 52 rounds to _____ tens.   J 19 rounds to _____ tens.

4

A   B   C   D   E

9⟌2 0 8 1   8⟌6 9 3 7   8⟌7 9 0 4   5⟌2 5 9 6   8⟌4 4 7 0

5

5)35	3)3	8)56	9)36	8)72	5)5	6)0
8)40	9)0	3)12	8)48	5)45	1)7	3)9
5)10	8)64	9)27	3)12	9)18	3)0	5)30
9)27	1)1	5)25	3)6	2)0	5)40	8)56
3)15	5)50	8)48	5)0	8)40	9)45	8)64
8)80	5)15	8)72	9)9	5)20		

6

8)48	2)8	2)4	8)40	2)6	8)64	2)6
8)56	2)8	8)72	2)4	8)80	2)6	2)8

7

Write the facts with no remainders.

A 8)77

B 8)60

C 8)44

D 8)83

E 8)53

F 8)59

8

A 3⟌1 1   B 8⟌4 5   C 3⟌3 5   D 8⟌5 9   E 5⟌4 8

9

A There are 2 paper clips in every box. There are 10 boxes. How many paper clips are there?

☐ _____

B Jill cooked dinner 8 times every month. She cooked dinner 40 times. How many months did Jill cook dinner?

☐ _____

C Each boat carries 5 people. There are 20 people in boats. How many boats are there?

☐ _____

Facts + Problems + Bonus = TOTAL

1

A 2⟌6

B 2⟌1 0

C 2⟌4

D 2⟌1 2

E 2⟌2

F 2⟌8

2

A 82 rounds to _____ tens.

B 33 rounds to _____ tens.

C 44 rounds to _____ tens.

D 59 rounds to _____ tens.

E 97 rounds to _____ tens.

F 74 rounds to _____ tens.

G 26 rounds to _____ tens.

H 91 rounds to _____ tens.

I 12 rounds to _____ tens.

J 68 rounds to _____ tens.

3

A 64 rounds to _____ tens.

B 96 rounds to _____ tens.

C 84 rounds to _____ tens.

D 26 rounds to _____ tens.

E 24 rounds to _____ tens.

F 36 rounds to _____ tens.

G 14 rounds to _____ tens.

H 16 rounds to _____ tens.

I 54 rounds to _____ tens.

J 56 rounds to _____ tens.

4

```
      A    6          B  1         C              D            E
    8 4 8 6 7        5 5 1 4      3 1 5 1 3      9 9 4 0 5      8 5 6 7 4
    - 4 8
        0
```

5

```
9 9        8 7 2      5 4 0      3 3        8 5 6      5 3 5      9 4 5

5 1 5      8 5 6      3 1 5      9 3 6      3 0        9 2 7      5 5

8 7 2      3 1 2      8 6 4      5 2 0      8 4 0      5 5 0      8 8 0

3 1 2      5 0        1 8        8 4 8      9 1 8      3 9        9 2 7

5 4 5      8 4 0      5 2 5      9 3 6      5 1 0      8 6 4      3 6

8 4 8      3 9        9 0        5 3 0      8 8 0
```

6

```
2 8        2 2        2 1 0      2 4        8 6 4      2 1 2      2 1 0
```

Part 6 continues on the next page.

2⟌6   8⟌4 0   2⟌1 2   8⟌5 6   2⟌6   2⟌4   8⟌8 0

2⟌1 2   8⟌4 8   2⟌1 0   8⟌7 2   2⟌8   8⟌3 2   2⟌2

7

Write the facts with no remainders.

A   8⟌7 3

B   8⟌5 1

C   8⟌4 3

D   8⟌8 7

E   8⟌6 2

F   8⟌6 8

8

A There are 9 apples in every bag. There are 18 apples in bags. How many bags of apples are there?

B Every hour Eric made 3 telephone calls. He made 15 calls. How many hours did Eric make calls?

C Ming taught 4 classes every year. He taught for 4 years. How many classes did Ming teach?

D Sue went fishing 9 times every year. She went fishing 36 times. How many years did Sue go fishing?

Part 8 continues on the next page.

E Marcy slept 4 hours each night. Eight
nights went by. How many hours did
Marcy sleep?

```
┌ ─ ─ ┐
│     │    _ _ _ _ _ _ _ _ _ _ _ _ _ _ _ _ _ _
└ ─ ─ ┘
```

F Each truck has 8 wheels. There are 48
wheels. How many trucks are there?

```
┌ ─ ─ ┐
│     │    _ _ _ _ _ _ _ _ _ _ _ _ _ _ _ _ _ _
└ ─ ─ ┘
```

9

Work the problems and find the remainders.

A

$8\overline{)529}$

B

$3\overline{)740}$

C

$9\overline{)407}$

D

$5\overline{)932}$

E

$9\overline{)318}$

F

$5\overline{)482}$

G

$3\overline{)172}$

H

$5\overline{)391}$

Lesson 30

Test + Facts + Bonus = TOTAL

1

A
2⟌1 6

B
2⟌2 0

C
2⟌1 8

D
2⟌1 4

2

A
2⟌1 4

B
2⟌1 6

C
2⟌1 8

D
2⟌2 0

3

A   B   C   D

4

A   87 rounds to _____ tens.

B   52 rounds to _____ tens.

C   57 rounds to _____ tens.

D   32 rounds to _____ tens.

E   17 rounds to _____ tens.

F   12 rounds to _____ tens.

G   77 rounds to _____ tens.

H   82 rounds to _____ tens.

5

A
```
    4
9|3 6 2 4
 -3 6
    0
```

B
```
    5
8|4 0 6 5
 -4 0
    0
```

C
5|4 5 3 0

D
3|1 2 1 0

E
9|1 8 3 8

6

A

9$\overline{)1\,2\,7\,0}$

B

8$\overline{)7\,6\,6\,4}$

C

3$\overline{)1\,3\,0\,4}$

D

5$\overline{)7\,2\,4\,0}$

E

9$\overline{)3\,7\,1}$

F

8$\overline{)4\,4\,6\,6}$

7

8$\overline{)4\,0}$ 5$\overline{)1\,5}$ 3$\overline{)0}$ 2$\overline{)1\,0}$ 3$\overline{)1\,2}$ 9$\overline{)2\,7}$ 5$\overline{)3\,0}$

3$\overline{)6}$ 2$\overline{)2}$ 9$\overline{)4\,5}$ 8$\overline{)8\,0}$ 5$\overline{)1\,0}$ 8$\overline{)5\,6}$ 5$\overline{)4\,5}$

9$\overline{)0}$ 5$\overline{)3\,5}$ 8$\overline{)4\,8}$ 2$\overline{)8}$ 2$\overline{)4}$ 5$\overline{)2\,0}$ 3$\overline{)3}$

Part 7 continues on the next page.

2) 1 2   3) 9   5) 5   9) 1 8   5) 5 0   2) 0   5) 4 0

9) 9   5) 0   2) 8   4) 0   8) 7 2   3) 1 2   5) 2 5

2) 6   3) 1 5   1) 5   8) 6 4   9) 3 6

8

2) 1 0   2) 1 4   2) 1 2   8) 4 8   2) 1 6   8) 7 2   2) 6

2) 1 4   2) 1 6   2) 8   8) 5 6   2) 1 2   2) 1 6   2) 1 4

Facts + Problems + Bonus = TOTAL

1

A
2⟌1 8

B
2⟌1 4

C
2⟌2 0

D
2⟌1 6

2

A
2⟌1 4

B
2⟌1 6

C
2⟌1 8

D
2⟌2 0

3

A
$$\begin{array}{r} 2 \\ 3\overline{)6\ 1} \\ -6 \\ \hline 0 \end{array}$$

B
$$\begin{array}{r} 1 \\ 9\overline{)9\ 7} \\ -9 \\ \hline 0 \end{array}$$

C
3⟌9 2

D
5⟌1 5 7

E
8⟌8 6

F
9⟌1 8 2

4

3⟌1 5   9⟌2 7   2⟌4   5⟌2 5   2⟌8   8⟌5 6   3⟌6

5⟌4 0   5⟌1 0   8⟌4 8   9⟌3 6   3⟌9   5⟌4 5   2⟌1 0

3⟌1 2   5⟌5   9⟌4 5   2⟌2   5⟌2 0   8⟌4 0   9⟌1 8

3⟌3   5⟌3 5   8⟌7 2   2⟌1 2   8⟌5 6   3⟌1 2   5⟌5 0

Part 4 continues on the next page.

$2\overline{)6}$ $5\overline{)30}$ $8\overline{)80}$ $5\overline{)35}$ $2\overline{)8}$ $5\overline{)15}$ $8\overline{)64}$

$9\overline{)9}$ $8\overline{)72}$ $5\overline{)30}$ $2\overline{)12}$ $3\overline{)9}$

5

$8\overline{)16}$ $2\overline{)14}$ $2\overline{)8}$ $2\overline{)12}$ $8\overline{)56}$ $2\overline{)16}$ $8\overline{)40}$

$8\overline{)64}$ $2\overline{)10}$ $2\overline{)14}$ $8\overline{)48}$ $2\overline{)16}$ $8\overline{)72}$ $2\overline{)14}$

6

Write the facts with no remainders.

A $2\overline{)5}$

B $2\overline{)13}$

C $2\overline{)9}$

D $2\overline{)7}$

E $2\overline{)3}$

F $2\overline{)11}$

7

A $9\overline{)3720}$

B $3\overline{)613}$

C $5\overline{)8290}$

D $8\overline{)4864}$

A Jack worked 8 hours every day. Jack worked for 48 hours. How many days did Jack work?

B There are 2 liters of pop in each bottle. I have 10 bottles. How many liters of pop do I have?

C There are 2 flowers in each vase. There are 8 vases. How many flowers are there?

D Ana skates 8 kilometers every day. How many days will it take her to skate 120 kilometers?

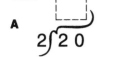

1

A 2⟌2 0

B 2⟌1 4

C 2⟌1 8

D 2⟌1 6

2

A 3 6 7   B 5 9 4   C 6 2 1   D 3 5 7

3

A
```
    4
5⟌2 0 3
 −2 0
     0
```

B
```
    5
7⟌3 5 6
 −3 5
     0
```

C 7⟌7 4

D 2⟌1 6 1

E 2⟌8 1

F 7⟌2 1 3

4

A
```
    2
3 4⟌9 9
    6 8
```

B
```
    2
3 1⟌7 5
```

C
```
    3
1 2⟌4 7
```

5

8⟌4 8   5⟌0   9⟌9   2⟌4   2⟌8   5⟌2 5   8⟌7 2

2⟌2   5⟌1 0   2⟌1 2   8⟌5 6   5⟌5 0   1⟌9   3⟌3

3⟌1 5   9⟌4 5   1⟌6   5⟌5   8⟌4 0   5⟌3 5   2⟌8

5⟌4 0   2⟌1 2   8⟌5 6   3⟌1 2   9⟌2 7   2⟌1 0   5⟌1 5

Part 5 continues on the next page.

$8\overline{)80}$ $9\overline{)18}$ $5\overline{)30}$ $2\overline{)6}$ $8\overline{)72}$ $3\overline{)9}$ $8\overline{)64}$

$3\overline{)6}$ $5\overline{)20}$ $9\overline{)36}$ $8\overline{)0}$ $5\overline{)45}$

6

$2\overline{)16}$ $2\overline{)10}$ $2\overline{)18}$ $2\overline{)8}$ $2\overline{)20}$ $2\overline{)14}$ $2\overline{)18}$

$8\overline{)64}$ $2\overline{)20}$ $2\overline{)16}$ $2\overline{)14}$ $2\overline{)18}$ $2\overline{)6}$ $2\overline{)20}$

$8\overline{)48}$ $2\overline{)4}$ $2\overline{)12}$

7

Write the facts with no remainders.

A $2\overline{)1}$

B $2\overline{)7}$

C $2\overline{)13}$

D $2\overline{)5}$

E $2\overline{)11}$

F $2\overline{)3}$

8

A Each bird has 4 worms. There are 12 birds. How many worms are there?

B There are 8 chairs for every table at the store. There are 48 chairs at the store. How many tables are there?

Part 8 continues on the next page.

c Leo has 9 records in each box. He has 27 records. How many boxes does he have?

d A store sells 4 shirts every week. The store sells shirts for 56 weeks. How many shirts did the store sell?

9

A

$5\overline{)1\ 5\ 2\ 5}$

B

$2\overline{)8\ 1\ 5}$

C

$5\overline{)8\ 3\ 2\ 5}$

D

$9\overline{)9\ 3\ 1}$

E

$2\overline{)1\ 3\ 6\ 8}$

Facts + Problems + Bonus = TOTAL

1

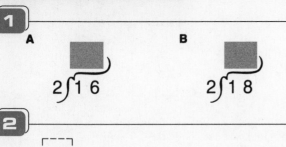

A 2⟌1 6

B 2⟌1 8

C 2⟌1 4

D 2⟌2 0

2

A 2⟌1 6

B 2⟌2 0

C 2⟌1 4

D 2⟌1 8

3

A 7⟌3 5

B 7⟌1 4

C 7⟌2 1

D 7⟌2 8

4

A 8 6 1   B 4 7 9   C 3 2 5   D 7 6 4

5

A
   3
21⟌7 9

B
   3
13⟌4 7 2

C
   1
42⟌5 8 3

D
   2
21⟌4 2

E
   4
32⟌1 4 9

F
   1
24⟌3 6 8

G
   4
31⟌1 3 7

H
   2
14⟌3 8

6

A Kathy drills 75 holes. She drills 5 holes every minute. How many minutes does it take her to drill the holes?

B A teacher bought 8 bags. Each bag has 214 peanuts. How many peanuts did the teacher buy?

C First Street Beach is 8 blocks long. There are 624 people on the beach. How many people are on each block of the beach?

D Shirley rode her bike in 8 races. Each race was 328 meters long. How many meters did Shirley ride her bike during the races?

7

8)56	5)20	2)6	9)27	2)4	5)5	8)40
5)45	1)3	9)18	2)2	5)50	3)9	3)0
8)48	2)12	5)15	9)9	8)64	2)4	5)35
9)0	2)10	8)72	5)40	1)10	3)15	2)8
5)25	3)3	5)20	2)12	8)80	9)36	5)10
3)6	5)30	3)12	9)45	5)25		

8

$2\overline{)18}$ $7\overline{)21}$ $7\overline{)14}$ $2\overline{)18}$ $7\overline{)21}$ $2\overline{)20}$ $7\overline{)14}$

$2\overline{)14}$ $2\overline{)20}$ $2\overline{)16}$ $7\overline{)21}$ $2\overline{)20}$ $7\overline{)14}$ $2\overline{)18}$

9

A	B	C	D
$2\overline{)1210}$	$8\overline{)562}$	$9\overline{)3627}$	$3\overline{)61}$

E	F	G	H
$2\overline{)7352}$	$8\overline{)62}$	$9\overline{)458}$	$9\overline{)31}$

I	J	K	L
$2\overline{)1387}$	$8\overline{)642}$	$8\overline{)408}$	$5\overline{)3254}$

Facts + Problems + Bonus = TOTAL

1

A 7⟌2 8

B 7⟌1 4

C 7⟌3 5

D 7⟌2 1

E 7⟌7

2

A 7⟌7

B 7⟌1 4

C 7⟌2 1

D 7⟌2 8

E 7⟌3 5

3

A   B   C   D

4

A 1 4 6   B 3 9 1   C 6 0 5   D 9 9 2

5

A 798 rounds to _____ tens.

B 474 rounds to _____ tens.

C 203 rounds to _____ tens.

D 645 rounds to _____ tens.

E 155 rounds to _____ tens.

F 297 rounds to _____ tens.

G 579 rounds to _____ tens.

H 512 rounds to _____ tens.

6

A $\frac{3}{3\,2\,\lfloor 9\,8\,5}$

B $\frac{1}{7\,3\,\lfloor 8\,3}$

C $\frac{6}{1\,0\,\lfloor 6\,6\,4}$

D $\frac{4}{4\,1\,\lfloor 1\,9\,7}$

Part 6 continues on the next page.

E $23\overline{)904}$ quotient 3   F $63\overline{)199}$ quotient 3   G $14\overline{)325}$ quotient 2

7

$9\overline{)36}$	$2\overline{)20}$	$5\overline{)35}$	$3\overline{)3}$	$2\overline{)4}$	$5\overline{)15}$	$5\overline{)50}$
$8\overline{)48}$	$3\overline{)6}$	$8\overline{)56}$	$2\overline{)14}$	$5\overline{)30}$	$8\overline{)80}$	$9\overline{)18}$
$2\overline{)10}$	$3\overline{)9}$	$5\overline{)25}$	$2\overline{)16}$	$9\overline{)27}$	$2\overline{)2}$	$5\overline{)45}$
$8\overline{)40}$	$5\overline{)20}$	$2\overline{)16}$	$2\overline{)12}$	$2\overline{)4}$	$8\overline{)64}$	$5\overline{)10}$
$8\overline{)48}$	$2\overline{)8}$	$3\overline{)12}$	$5\overline{)40}$	$5\overline{)5}$	$3\overline{)15}$	$2\overline{)6}$
$9\overline{)45}$	$8\overline{)72}$	$2\overline{)8}$	$2\overline{)18}$	$9\overline{)9}$		

8

$2\overline{)10}$	$7\overline{)14}$	$2\overline{)4}$	$2\overline{)18}$	$2\overline{)12}$	$2\overline{)8}$	$7\overline{)21}$
$7\overline{)21}$	$2\overline{)2}$	$7\overline{)14}$	$2\overline{)16}$	$2\overline{)20}$	$7\overline{)21}$	$2\overline{)6}$

9

A	B	C	D
2$\overline{7\,3\,8\,5}$	8$\overline{3\,4\,4\,7}$	9$\overline{1\,8\,4\,3}$	5$\overline{3\,6\,6\,1}$

10

A There are 8 seats in each boat. There are 336 seats. How many boats are there?

B Randy corrected papers for 294 class periods. He corrected 7 papers every class period. How many papers did he correct?

C Every morning Jason writes 3 pages in his book. He writes for 78 mornings. How many pages long will his book be?

D Phil has 732 coins in a jar. Each day he puts 3 coins in the jar. How many days has Phil been saving coins?

Test + Facts + Problems + Bonus = TOTAL

1

A 7⟌2 1

B 7⟌2 8

C 7⟌7

D 7⟌3 5

E 7⟌1 4

2

A 265 rounds to _____ tens.

B 304 rounds to _____ tens.

C 737 rounds to _____ tens.

D 972 rounds to _____ tens.

E 180 rounds to _____ tens.

F 846 rounds to _____ tens.

G 591 rounds to _____ tens.

H 459 rounds to _____ tens.

3

A $\overset{2\quad3}{18\overline{)65}}$

B $\overset{9}{23\overline{)210}}$

C $\overset{6}{34\overline{)228}}$

D $\overset{3}{47\overline{)150}}$

E $\overset{6}{46\overline{)290}}$

4

A 2⟌8 1

B 2⟌7 3 5 8

C 8⟌4 0 1 8

D 7⟌9 1 4

E 5⟌3 5 8 5

5

5⟌2 5	8⟌4 8	2⟌6	1⟌4	9⟌3 6	3⟌1 2	5⟌1 0
5⟌1 5	3⟌3	2⟌0	9⟌1 8	8⟌5 6	5⟌4 5	2⟌2
2⟌2 0	5⟌4 0	2⟌4	8⟌5 6	3⟌9	1⟌8	5⟌3 0
8⟌4 0	5⟌2 0	3⟌1 5	5⟌5	2⟌1 0	9⟌9	5⟌5 0
9⟌2 7	2⟌8	6⟌0	8⟌8 0	2⟌1 6	5⟌3 5	2⟌1 2
8⟌7 2	8⟌4 8	8⟌6 4	3⟌6	9⟌4 5		

6

2⟌1 0	2⟌1 8	7⟌2 8	7⟌1 4	7⟌3 5	2⟌1 2	7⟌3 5
7⟌2 8	2⟌1 6	7⟌2 1	2⟌1 4	7⟌3 5	7⟌2 8	2⟌6

7

Write the facts with no remainders.

A   2⟌1 7

B   2⟌1 9

C   2⟌1 5

D   2⟌1 1

E   2⟌1 5

F   2⟌1 7

8

A Ed fried 36 eggs. He fried 2 eggs every morning. How many mornings did Ed fry eggs?

B Every time Dave bought stamps he wrote 2 letters. He bought stamps 32 times. How many letters did Dave write?

C There are 9 rows of plants. Each row has 173 plants. How many plants are there?

D There are 9 students in the blue group. Each student in the blue group earned 387 points. How many points did the students earn in all?

9

Multiply to find the number you subtract. Then find the first remainder.

A
$$72\overline{)3092}$$ 4

B
$$90\overline{)400}$$ 4

C
$$43\overline{)891}$$ 2

D
$$13\overline{)507}$$ 3

Facts + Problems + Bonus = TOTAL

1

A 4⟌2 8   B 4⟌4 0   C 4⟌3 2   D 4⟌3 6

2

A 4⟌2 4

B 4⟌2 8

C 4⟌3 2

D 4⟌3 6

E 4⟌4 0

3

A   B   C   D   E

```
     6          2          5          3          6
16 |9 7 1|  37 |9 5 8|  54 |3 0 0|  46 |1 5 7 2|  48 |3 2 1|
```

4

A 7⟌1 4

B 7⟌3 5

C 7⟌2 8

D 7⟌7

E 7⟌2 1

5

```
A        8       B       3       C         5      D        4      E        7
   41 ⌐378        18 ⌐72      24 ⌐138        46 ⌐233        38 ⌐287
     −328          −54          −120          −184          −266
       50           18           18            49            21
```

6

```
2 ⌐8      3 ⌐3      8 ⌐40      5 ⌐50      9 ⌐18      2 ⌐4      5 ⌐30

8 ⌐72     2 ⌐16     5 ⌐20      2 ⌐2       3 ⌐9       8 ⌐48     9 ⌐36

5 ⌐45     2 ⌐10     8 ⌐64      2 ⌐8       5 ⌐40      5 ⌐15     2 ⌐6

5 ⌐25     3 ⌐12     9 ⌐27      3 ⌐6       2 ⌐14      7 ⌐0      5 ⌐10

2 ⌐20     1 ⌐2      2 ⌐12      5 ⌐35      9 ⌐45      8 ⌐56     3 ⌐15

2 ⌐16     2 ⌐12     8 ⌐80      9 ⌐9       2 ⌐18
```

7

```
7 ⌐21     4 ⌐36     4 ⌐32      7 ⌐28      4 ⌐40      7 ⌐14     7 ⌐35

4 ⌐36     4 ⌐40     7 ⌐28      4 ⌐32      7 ⌐35      4 ⌐36     7 ⌐28
```

8

Write the facts with no remainders.

A   2 ⌐11

B   2 ⌐15

C   2 ⌐19

D   2 ⌐17

E   2 ⌐19

F   2 ⌐15

9

Multiply to find the number you subtract. Then find the first remainder.

A
$$\begin{array}{r} 8 \\ 10\overline{)8\ 2\ 5} \end{array}$$

B
$$\begin{array}{r} 2 \\ 24\overline{)5\ 8} \end{array}$$

C
$$\begin{array}{r} 2 \\ 84\overline{)1\ 9\ 8} \end{array}$$

D
$$\begin{array}{r} 3 \\ 31\overline{)9\ 8\ 5} \end{array}$$

10

A
$$7\overline{)7\ 2\ 8}$$

B
$$2\overline{)1\ 1\ 4\ 1}$$

C
$$5\overline{)2\ 7\ 6\ 5}$$

D
$$8\overline{)4\ 0\ 5\ 6}$$

11

A 407 rounds to _____ tens.

B 525 rounds to _____ tens.

C 649 rounds to _____ tens.

D 911 rounds to _____ tens.

E 397 rounds to _____ tens.

12

A Each team will have 9 children. There are 315 children who want to be on teams. How many teams will there be?

B The ABC Company is going to make 96 trucks. Each truck will have 8 tires. How many tires will the ABC Company need?

C There are 8 windows in each house. There are 124 houses. How many windows are there in all?

D Harry drank 5 glasses of water each day. He drank 150 glasses of water. How many days did he drink water?

Lesson 37

Facts + Problems + Bonus = TOTAL

1

A  1 0 ⟌5 0

B 1 0 ⟌8 0

C 1 0 ⟌4 0

D 1 0 ⟌7 0

2

A 1 0 ⟌1 0

B 1 0 ⟌2 0

C 1 0 ⟌3 0

D 1 0 ⟌4 0

E 1 0 ⟌5 0

F 1 0 ⟌6 0

G 1 0 ⟌7 0

H 1 0 ⟌8 0

I 1 0 ⟌9 0

J 1 0 ⟌1 0 0

3

A 4 ⟌3 2

B 4 ⟌2 4

C 4 ⟌3 6

D 4 ⟌2 8

4

A 4 ⟌2 4

B 4 ⟌2 8

C 4 ⟌3 2

D 4 ⟌3 6

E 4 ⟌4 0

86 Lesson 37

5

A
```
        6
1 4 | 9 8        ┌─────┐
  − 8 4          │     │
    1 4
```

B
```
        4
2 8 | 1 4 2      ┌─────┐
  − 1 1 2        │     │
      3 0
```

C
```
        5
3 3 | 1 9 1      ┌─────┐
  − 1 6 5        │     │
      2 6
```

D
```
        7
2 4 | 1 9 8      ┌─────┐
  − 1 6 8        │     │
      3 0
```

6

2 | 1 8   8 | 4 8   5 | 3 5   7 | 1 4   9 | 4 5   3 | 1 2   2 | 8

7 | 2 1   5 | 5   2 | 1 6   9 | 3 6   8 | 5 6   2 | 2   5 | 4 0

8 | 4 0   2 | 1 0   5 | 2 0   3 | 9   7 | 2 8   5 | 5 0   2 | 1 4

5 | 4 5   7 | 7   7 | 2 1   2 | 4   5 | 3 0   8 | 6 4   5 | 1 5

2 | 1 2   5 | 2 5   3 | 6   8 | 7 2   2 | 2 0   9 | 2 7   2 | 6

5 | 1 0   7 | 3 5   8 | 8 0   9 | 1 8   3 | 1 5

7

10 | 2 0   10 | 4 0   4 | 3 2   10 | 5 0   4 | 4 0   4 | 3 6   10 | 5 0

4 | 4 0   4 | 3 6   10 | 1 0   10 | 3 0   4 | 3 2   10 | 4 0   4 | 3 6

8

Write the facts with no remainders.

A ☐ 2)1 7 ⌐

B ☐ 2)1 5 ⌐

C ☐ 2)2 1 ⌐

D ☐ 2)1 9 ⌐

E ☐ 2)1 3 ⌐

F ☐ 2)1 1 ⌐

9

Multiply to find the number you subtract. Then find the first remainder.

A 3
2 6)1 0 1 4

B 4
4 3)1 7,2 5 9

C 3
3 2)1 1 4

D 2
2 9)6 7 4

E 2
1 4)3 4 9 5

10

A 108 rounds to _____ tens.

B 929 rounds to _____ tens.

C 860 rounds to _____ tens.

D 666 rounds to _____ tens.

E 345 rounds to _____ tens.

F 378 rounds to _____ tens.

11

A A teacher came to school for 120 days. She drove 5 kilometers every day. How many kilometers did she drive?

B 125 cars came into a parking lot. There were 5 people in each car. How many people came into the parking lot?

C A machine wrapped 9 loaves of bread every minute. It wrapped 1260 loaves of bread. How many minutes did it take to wrap the loaves?

D Ray uses 7 liters of paint each week. He has used 2128 liters. How many weeks has Ray painted?

Lesson 38

Facts + Problems + Bonus = TOTAL

1

A ▢) 10⟌30

B ▢) 10⟌60

C ▢) 10⟌90

D ▢) 10⟌100

2

A ⬚⟌ 10⟌10

B ⬚⟌ 10⟌20

C ⬚⟌ 10⟌30

D ⬚⟌ 10⟌40

E ⬚⟌ 10⟌50

F ⬚⟌ 10⟌60

G ⬚⟌ 10⟌70

H ⬚⟌ 10⟌80

I ⬚⟌ 10⟌90

J ⬚⟌ 10⟌100

3

A
```
      4
3 1 ⟌1 6 9
   -1 2 4
      4 5
```

B
```
      2
1 4 ⟌4 2
   - 2 8
     1 4
```

C
```
      6
2 3 ⟌1 5 7
   -1 3 8
      1 9
```

D
```
      3
4 5 ⟌1 8 2
   -1 3 5
      4 7
```

4

A   4⟌2 8

B   4⟌3 2

C   4⟌4 0

D   4⟌3 6

E   4⟌2 4

5

2⟌1 6   8⟌4 0   2⟌4   3⟌1 5   9⟌2 7   7⟌1 4   5⟌2 0

8⟌8 0   3⟌1 2   5⟌4 5   7⟌2 8   5⟌3 5   2⟌2   8⟌5 6

2⟌6   7⟌7   5⟌1 5   2⟌1 4   8⟌4 8   5⟌3 0   2⟌1 0

9⟌1 8   5⟌1 0   3⟌9   8⟌7 2   7⟌3 5   2⟌2 0   9⟌3 6

7⟌2 1   5⟌4 0   2⟌1 8   9⟌4 5   2⟌8   3⟌6   5⟌2 5

2⟌1 2   9⟌9   3⟌3   5⟌5 0   8⟌6 4

6

4⟌3 6   10⟌3 0   4⟌2 8   4⟌4 0   10⟌5 0   4⟌2 4   10⟌2 0

10⟌4 0   4⟌2 4   4⟌3 2   10⟌3 0   4⟌2 8   10⟌5 0   4⟌3 6

4⟌2 8   10⟌2 0   4⟌4 0   10⟌1 0 0   4⟌3 2   10⟌1 0   4⟌0

7

A 271 rounds to _____ tens.　　　B 350 rounds to _____ tens.

C 888 rounds to _____ tens.　　　D 695 rounds to _____ tens.

E 704 rounds to _____ tens.

8

Write the facts with no remainders.

A

7)31 ⌐

B

7)24 ⌐

C

7)5 ⌐

D

7)18 ⌐

E

7)10 ⌐

F

7)38 ⌐

9

Multiply to find the number you subtract. Then find the first remainder.

A   4
24)1 0 3 2

B   5
53)2 8 6 4

C   2
27)5 9 3 2

D   5
42)2 3 0 8

E   5
37)1 8 9

10

A   B   C   D   E

3)7 2 1   5)5 7 0   8)7 0 0 9   7)7 3 5   2)6 1

A A machine has been making boxes for 7 hours. The machine makes 214 boxes each hour. How many boxes has the machine made?

B Each time Ella fixes a suit, she uses 7 pins. Ella has used 84 pins. How many suits has Ella fixed?

C A factory makes 9 cars every hour. The factory made 1080 cars. How many hours did it take to make the cars?

D Each student collected 7 cans. There are 1425 students in our school. How many cans were collected in all?

1

A
$$10\overline{)60}$$

B
$$10\overline{)40}$$

C
$$10\overline{)100}$$

D
$$10\overline{)30}$$

E
$$10\overline{)70}$$

F
$$10\overline{)50}$$

G
$$10\overline{)90}$$

H
$$10\overline{)10}$$

I
$$10\overline{)80}$$

J
$$10\overline{)20}$$

2

A
```
      3
62 | 2 5 6
   -1 8 6
      7 0
```

B
```
      6
25 | 1 7 0
   -1 5 0
      2 0
```

C
```
      5
43 | 2 6 2
   -2 1 5
      4 7
```

D
```
      3
38 | 1 5 2
   -1 1 4
      3 8
```

3

A
```
      8
41 | 3 5 0
   -3 2 8
```

B
```
      5
26 | 1 4 2
   -1 3 0
```

Part 3 continues on the next page.

C

```
            6
  3 7 | 2 1 8
    − 2 2 2
```

D

```
        4
  1 8 | 7 0
    − 7 2
```

4

A

```
  4 ) 3 2
```

B

```
  4 ) 3 6
```

C

```
  4 ) 2 4
```

D

```
  4 ) 4 0
```

E

```
  4 ) 2 8
```

5

A

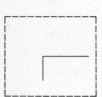

```
            6
  5 4 | 3 2 6
```

```
      6
  5 ) 3 3
```

B

```
  2 9 | 7 3 9
```

C

```
  3 3 | 2 0 2 1
```

D

```
  3 2 | 5 0 5
```

E

```
  1 8 | 3 2 8
```

6

A 704 rounds to _____ tens.

B 919 rounds to _____ tens.

C 592 rounds to _____ tens.

D 495 rounds to _____ tens.

E 655 rounds to _____ tens.

F 837 rounds to _____ tens.

7

7⟌28	9⟌36	2⟌20	5⟌35	3⟌3	2⟌10	5⟌15
3⟌12	2⟌4	7⟌7	8⟌40	9⟌18	5⟌45	2⟌12
2⟌16	8⟌72	5⟌25	2⟌2	5⟌10	7⟌35	5⟌50
5⟌30	9⟌27	2⟌18	7⟌14	8⟌48	2⟌8	5⟌20
7⟌21	8⟌80	2⟌6	8⟌56	3⟌9	9⟌45	5⟌40
8⟌64	3⟌15	3⟌6	9⟌9	2⟌14		

8

10⟌40	4⟌28	10⟌70	4⟌36	10⟌90	4⟌24	10⟌20
10⟌100	4⟌24	10⟌10	4⟌40	4⟌32	10⟌70	10⟌100
4⟌28	10⟌80	10⟌60	4⟌28	10⟌30	10⟌80	4⟌32

9

Write the facts with no remainders.

A   7⟌13

B   7⟌39

C   7⟌9

D   7⟌30

E   7⟌23

F   7⟌19

10

A $8\overline{)643}$ B $2\overline{)1375}$ C $9\overline{)927}$ D $5\overline{)2853}$ E $7\overline{)3524}$

11

Multiply to find the number you subtract. Then find the first remainder.

A $\overset{6}{24\overline{)1582}}$ B $\overset{5}{16\overline{)8243}}$ C $\overset{2}{46\overline{)983}}$ D $\overset{5}{38\overline{)2039}}$

12

A Our family eats 3 cans of beans every week. We have 72 cans of beans. How many weeks will the beans last?

B Alex made 3 rings every hour. He made 1320 rings. How many hours did Alex make rings?

C The pilot practiced for 3 hours every day. She practiced for 75 hours. How many days did the pilot practice?

D Lola used 5 batteries every month. She used batteries for 315 months. How many batteries did Lola use?

Lesson 40

Facts + Problems + Bonus = TOTAL

1

A 6 ⟌ 2 4

B 6 ⟌ 1 2

C 6 ⟌ 1 8

D 6 ⟌ 3 0

2

A 6 ⟌ 6

B 6 ⟌ 1 2

C 6 ⟌ 1 8

D 6 ⟌ 2 4

E 6 ⟌ 3 0

3

A   B   C   D

4

A 1 0 ⟌ 9 0

B 1 0 ⟌ 1 0

C 1 0 ⟌ 5 0

D 1 0 ⟌ 2 0

E 1 0 ⟌ 8 0

F 1 0 ⟌ 3 0

G 1 0 ⟌ 7 0

H 1 0 ⟌ 4 0

I 1 0 ⟌ 1 0 0

J 1 0 ⟌ 6 0

Lesson 40 —— ■ **97**

5

A
$$
\begin{array}{r}
5 \\
29\overline{)178} \\
-145 \\
\hline
33
\end{array}
$$

B
$$
\begin{array}{r}
4 \\
35\overline{)168} \\
-140 \\
\hline
28
\end{array}
$$

C
$$
\begin{array}{r}
5 \\
25\overline{)150} \\
-125 \\
\hline
25
\end{array}
$$

6

A
$$
\begin{array}{r}
8 \\
46\overline{)374} \\
-368 \\
\hline
\end{array}
$$

B
$$
\begin{array}{r}
6 \\
24\overline{)167} \\
-144 \\
\hline
\end{array}
$$

C
$$
\begin{array}{r}
5 \\
36\overline{)168} \\
-180 \\
\hline
\end{array}
$$

D
$$
\begin{array}{r}
9 \\
31\overline{)274} \\
-279 \\
\hline
\end{array}
$$

7

A $87\overline{)391}$

B $54\overline{)3828}$

C $26\overline{)847}$

D $75\overline{)6982}$

8

$4\overline{)32}$ $8\overline{)56}$ $7\overline{)28}$ $5\overline{)40}$ $2\overline{)18}$ $5\overline{)35}$ $9\overline{)27}$

$2\overline{)12}$ $3\overline{)15}$ $8\overline{)48}$ $4\overline{)24}$ $9\overline{)36}$ $2\overline{)6}$ $5\overline{)25}$

Part 8 continues on the next page.

$9\overline{)45}$ $2\overline{)16}$ $3\overline{)9}$ $2\overline{)20}$ $7\overline{)21}$ $8\overline{)64}$ $4\overline{)36}$

$8\overline{)40}$ $4\overline{)28}$ $5\overline{)10}$ $2\overline{)4}$ $9\overline{)18}$ $3\overline{)12}$ $8\overline{)72}$

$2\overline{)14}$ $3\overline{)6}$ $7\overline{)14}$ $5\overline{)15}$ $5\overline{)30}$ $4\overline{)40}$ $5\overline{)20}$

$2\overline{)8}$ $7\overline{)35}$ $8\overline{)80}$ $2\overline{)10}$ $5\overline{)45}$

$10\overline{)90}$ $6\overline{)12}$ $10\overline{)60}$ $10\overline{)20}$ $6\overline{)18}$ $10\overline{)70}$ $10\overline{)50}$

$10\overline{)100}$ $6\overline{)18}$ $10\overline{)80}$ $6\overline{)12}$ $10\overline{)90}$ $10\overline{)60}$ $10\overline{)40}$

$10\overline{)10}$ $10\overline{)100}$ $6\overline{)18}$ $10\overline{)70}$ $6\overline{)12}$ $10\overline{)30}$ $10\overline{)80}$

A 233 rounds to _____ tens.   B 116 rounds to _____ tens.

C 384 rounds to _____ tens.   D 808 rounds to _____ tens.

E 195 rounds to _____ tens.   F 497 rounds to _____ tens.

Write the facts with no remainders.

A
$7\overline{)17}$

B
$7\overline{)33}$

C
$7\overline{)26}$

D $7\overline{)12}$

E $7\overline{)3}$

F
$7\overline{)37}$

12

A   B   C   D

$5\overline{)1302}$ $2\overline{)801}$ $3\overline{)423}$ $5\overline{)7524}$

13

Multiply to find the number you subtract. Then find the first remainder.

A
$$17\overline{)3894} \quad ^2$$

B
$$52\overline{)248} \quad ^4$$

C
$$24\overline{)9853} \quad ^4$$

D
$$83\overline{)1452} \quad ^1$$

14

A Each watch needed 4 wheels. There were 412 wheels. How many watches could be made?

B Mr. Singu worked 1232 hours. He made 4 toys each hour he worked. How many toys did Mr. Singu make?

C There were 7 men connecting wires. 308 wires were connected. How many wires did each man connect?

D Each pie needed 15 apples. Amy made 3 pies. How many apples did Amy use?

Lesson 41

1

A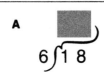
6⟌1 8

B
6⟌3 0

C
6⟌1 2

D
6⟌2 4

2

A
6⟌6

B
6⟌1 2

C
6⟌1 8

D
6⟌2 4

E
6⟌3 0

3

A
2
6⟌1 2

B
3
4⟌1 2

4

A   B   C   D

5

A
    4
3 7⟌1 3 2
  −1 4 8

B
    5
4 1⟌2 1 0
  −2 0 5

C
    2
2 4⟌3 8
  −4 8

D
    6
2 9⟌1 7 0
  −1 7 4

6

A

$$
\begin{array}{r} 9 \\ 83\overline{)759} \end{array}
$$

B

$$
\begin{array}{r} 7 \\ 62\overline{)499} \end{array}
$$

C

$$
\begin{array}{r} 8 \\ 58\overline{)452} \end{array}
$$

D

$$
\begin{array}{r} 8 \\ 71\overline{)640} \end{array}
$$

7

A

$$41\overline{)3129}$$

B

$$19\overline{)1331}$$

C

$$85\overline{)245}$$

D

$$67\overline{)2836}$$

E

$$54\overline{)681}$$

F

$$81\overline{)4444}$$

G

$$38\overline{)3527}$$

8

$$4\overline{)32} \qquad 2\overline{)18} \qquad 10\overline{)80} \qquad 7\overline{)28} \qquad 2\overline{)16} \qquad 10\overline{)100} \qquad 8\overline{)64}$$

$$10\overline{)20} \qquad 4\overline{)28} \qquad 8\overline{)48} \qquad 9\overline{)45} \qquad 10\overline{)40} \qquad 7\overline{)21} \qquad 4\overline{)36}$$

Part 8 continues on the next page.

$5\overline{)45}$ $10\overline{)10}$ $8\overline{)40}$ $2\overline{)12}$ $10\overline{)60}$ $9\overline{)18}$ $5\overline{)40}$

$9\overline{)36}$ $10\overline{)50}$ $5\overline{)35}$ $7\overline{)14}$ $2\overline{)14}$ $2\overline{)8}$ $10\overline{)70}$

$5\overline{)20}$ $9\overline{)27}$ $10\overline{)30}$ $4\overline{)24}$ $7\overline{)35}$ $10\overline{)90}$ $8\overline{)72}$

$5\overline{)25}$ $8\overline{)56}$ $4\overline{)32}$ $5\overline{)30}$ $4\overline{)40}$

9

$10\overline{)30}$ $6\overline{)18}$ $10\overline{)70}$ $10\overline{)20}$ $6\overline{)12}$ $6\overline{)18}$ $10\overline{)100}$

$6\overline{)18}$ $10\overline{)80}$ $10\overline{)10}$ $6\overline{)12}$ $10\overline{)60}$ $10\overline{)90}$ $10\overline{)40}$

10

Write the facts with no remainders.

A $4\overline{)30}$ **B** $4\overline{)41}$ **C** $4\overline{)26}$

D $4\overline{)33}$ **E** $4\overline{)35}$ **F** $4\overline{)38}$

11

$9\overline{)1845}$ $4\overline{)3921}$ $3\overline{)1056}$ $8\overline{)5367}$ $7\overline{)7211}$

12

A Mrs. Smith bought 7 rolls of string. She can tie 84 packages with each roll. How many packages can she tie with the string she bought?

B 4 women are holding each rope. There are 36 women in all. How many ropes are there?

C Jim does 2 sit-ups whenever he exercises. He has done 90 sit-ups. How many times has Jim exercised?

13

Multiply to find the number you subtract. Then find the remainder.

A
$$\begin{array}{r} 3 \\ 2\,4\,\overline{\smash)9\,1\,8\,5} \end{array}$$

B
$$\begin{array}{r} 5 \\ 8\,2\,\overline{\smash)4\,2\,3\,8} \end{array}$$

C
$$\begin{array}{r} 4 \\ 2\,5\,\overline{\smash)1\,0\,8} \end{array}$$

D
$$\begin{array}{r} 4 \\ 3\,7\,\overline{\smash)1\,8\,4\,9} \end{array}$$

Lesson 42

1

A 6⟌1 8

B 6⟌2 4

C 6⟌1 2

D 6⟌3 0

E 6⟌6

2

A 2⟍ 9⟌1 8

B 6⟍ 3⟌1 8

3

A 7 6 5⟌5 0 5

B 2 9 6⟌3 0 4

C 8 5 8⟌4 3 1

D 4 3 7⟌1 2 9

E 5 4 6⟌2 7 6

F 6 5 2⟌2 9 8

4

A 2 1⟌8 6 8

B 3 3⟌7 4 3

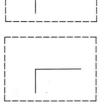

5

4⟌32	10⟌100	7⟌14	5⟌25	10⟌30	8⟌40	9⟌45
2⟌14	9⟌18	10⟌80	3⟌15	2⟌6	8⟌64	5⟌20
7⟌35	10⟌50	2⟌16	8⟌56	10⟌20	5⟌35	7⟌21
4⟌24	2⟌12	5⟌40	9⟌27	4⟌36	3⟌12	10⟌90
2⟌8	10⟌60	9⟌36	4⟌28	8⟌72	10⟌40	7⟌28
2⟌18	8⟌48	3⟌9	10⟌70	2⟌10		

6

6⟌24	4⟌32	6⟌18	6⟌30	4⟌24	6⟌12	6⟌30
4⟌28	6⟌12	10⟌30	10⟌70	6⟌24	4⟌36	6⟌18

7

A Fran brought 2 worms every time she went fishing. She went fishing 134 times. How many worms did Fran bring?

⌐ ‾ ‾ ‾ ┐
└ _ _ _ ┘   _ _ _ _ _ _ _ _ _ _ _ _

B Tina steered the ship 4 hours a day. She steered the ship for 120 hours. How many days did Tina steer the ship?

⌐ ‾ ‾ ‾ ┐
└ _ _ _ ┘   _ _ _ _ _ _ _ _ _ _ _ _

Part 7 continues on the next page.

Lesson 42 (continued)

c Every zoo I visited had 7 lions. I saw 140 lions. How many zoos did I visit?

d Every time Leroy went skiing, he went down 9 hills. He went skiing 216 times. How many hills did Leroy go down?

8

A 4)2830   **B** 7)861   **C** 2)1994   **D** 8)5672   **E** 5)1600

9

Write the facts with no remainders.

A 4)37

B 4)39

C 4)27

D 4)29

E 4)42

F 4)34

G 10)46

H 10)21

I 10)85

J 10)97

K 10)32

L 10)104

10

Multiply to find the number you subtract. Then find the remainder.

A
         2
3 6 | 9 2 4 5

B
         3
5 8 | 1 8 3 6

C
         1
1 9 | 2 1 3 4

D
         4
4 2 | 2 0 0 0

Lesson 43

Facts	+	Problems	+	Bonus	=	TOTAL

1

A $7\overline{)56}$

B $7\overline{)70}$

C $7\overline{)63}$

D $7\overline{)42}$

2

A $7\overline{)42}$

B $7\overline{)49}$

C $7\overline{)56}$

D $7\overline{)63}$

E $7\overline{)70}$

3

A $38\overline{)186}$ 5

B $52\overline{)275}$ 4

C $46\overline{)129}$ 3

D $26\overline{)161}$ 7

E $46\overline{)277}$ 5

4

A $6\overline{)30}$

B $6\overline{)6}$

C $6\overline{)24}$

D $6\overline{)18}$

E $6\overline{)12}$

5

A

B

$$7\,7\,\overline{)4\,4\,9\,2}$$

$$8\,6\,\overline{)2\,8\,2\,7}$$

6

$4\overline{)24}$	$8\overline{)56}$	$7\overline{)28}$	$5\overline{)10}$	$1\overline{)9}$	$2\overline{)16}$	$3\overline{)9}$
$5\overline{)30}$	$10\overline{)40}$	$9\overline{)27}$	$8\overline{)48}$	$5\overline{)45}$	$4\overline{)28}$	$5\overline{)50}$
$8\overline{)64}$	$8\overline{)0}$	$4\overline{)36}$	$3\overline{)12}$	$10\overline{)50}$	$2\overline{)18}$	$9\overline{)36}$
$3\overline{)15}$	$2\overline{)8}$	$9\overline{)18}$	$4\overline{)32}$	$4\overline{)40}$	$7\overline{)35}$	$3\overline{)6}$
$7\overline{)21}$	$5\overline{)15}$	$10\overline{)60}$	$8\overline{)40}$	$2\overline{)14}$	$9\overline{)45}$	$1\overline{)3}$
$4\overline{)0}$	$2\overline{)10}$	$10\overline{)10}$	$8\overline{)72}$	$2\overline{)20}$		

7

$6\overline{)30}$	$7\overline{)42}$	$6\overline{)24}$	$7\overline{)49}$	$6\overline{)18}$	$7\overline{)42}$	$6\overline{)30}$
$6\overline{)24}$	$6\overline{)18}$	$7\overline{)49}$	$7\overline{)42}$	$6\overline{)12}$	$6\overline{)30}$	$6\overline{)24}$

8

Write the facts with no remainders.

A $4\overline{)43}$

B $4\overline{)31}$

C $4\overline{)25}$

D $4\overline{)38}$

E $4\overline{)34}$

F $4\overline{)30}$

9

A It rained 5 centimeters of water every hour. It rained for 80 hours. How many centimeters of water did it rain?

B 4 people can sit at each table. There are 420 people sitting at tables. How many tables are there?

C I have 7 dishes on each shelf. If I have 140 dishes, how many shelves do I have?

10

A $2\overline{)9601}$

B $4\overline{)2435}$

C $2\overline{)2081}$

D $7\overline{)2943}$

E $5\overline{)1984}$

11

Multiply to find the number you subtract. Then find the remainder.

A $\begin{array}{r} 8 \\ 15\overline{)1349} \end{array}$

B $\begin{array}{r} 4 \\ 20\overline{)8372} \end{array}$

C $\begin{array}{r} 2 \\ 47\overline{)985} \end{array}$

D $\begin{array}{r} 3 \\ 42\overline{)1354} \end{array}$

Lesson 44

Facts + Problems + Bonus = TOTAL

1

A 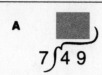 7⟌4 9

B 7⟌6 3

C 7⟌5 6

D 7⟌4 2

2

A 7⟌4 2

B 7⟌4 9

C 7⟌5 6

D 7⟌6 3

E 7⟌7 0

3

A   B   C   D

4

A
```
      4
76⟌3 9 3
```

B
```
       6
28⟌1 5 3
```

C
```
       3
47⟌1 3 0
```

D
```
       4
65⟌3 2 5
```

E
```
       8
33⟌2 4 7
```

F
```
       5
46⟌2 7 7
```

5

A

$58\overline{)1872}$

B

$72\overline{)984}$

C

$48\overline{)948}$

D

$62\overline{)2561}$

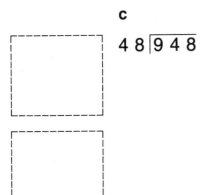

6

$6\overline{)30}$ $9\overline{)18}$ $7\overline{)28}$ $3\overline{)15}$ $4\overline{)28}$ $8\overline{)48}$ $7\overline{)7}$

$5\overline{)5}$ $2\overline{)12}$ $4\overline{)24}$ $8\overline{)56}$ $6\overline{)12}$ $3\overline{)9}$ $9\overline{)45}$

$5\overline{)35}$ $6\overline{)6}$ $2\overline{)4}$ $7\overline{)21}$ $9\overline{)36}$ $5\overline{)25}$ $4\overline{)40}$

$3\overline{)12}$ $4\overline{)32}$ $9\overline{)27}$ $6\overline{)18}$ $2\overline{)14}$ $8\overline{)80}$ $7\overline{)35}$

$10\overline{)20}$ $8\overline{)72}$ $7\overline{)14}$ $8\overline{)40}$ $5\overline{)40}$ $2\overline{)6}$ $6\overline{)24}$

$5\overline{)20}$ $10\overline{)100}$ $8\overline{)64}$ $3\overline{)6}$ $4\overline{)36}$

7

$7\overline{)49}$ $6\overline{)18}$ $7\overline{)42}$ $4\overline{)28}$ $6\overline{)6}$ $7\overline{)49}$ $4\overline{)32}$

$7\overline{)42}$ $6\overline{)12}$ $4\overline{)24}$ $7\overline{)49}$ $6\overline{)30}$ $7\overline{)42}$ $4\overline{)36}$

8

A Mr. and Mrs. Kato danced 8 dances every night. They danced 280 dances. How many nights did Mr. and Mrs. Kato go dancing?

B A squirrel cracked 8 nuts every minute. It cracked nuts for 40 minutes. How many nuts did the squirrel crack?

C Beth took 6 pictures every hour. She took 192 pictures. How many hours did Beth take pictures?

9

A   B   C   D

$6\overline{)8523}$ $7\overline{)1490}$ $5\overline{)2300}$ $4\overline{)2436}$

Facts + Problems + Bonus = TOTAL

1

A 4⟌8   B 4⟌2 0   C 4⟌1 2   D 4⟌1 6

2

A 4⟌4   B 4⟌8

C 4⟌1 2   D 4⟌1 6

E 4⟌2 0

3

A 10⟌   B 5⟌
2⟌2 0   4⟌2 0

4

A 7⟌5 6   B 7⟌6 3

C 7⟌4 9   D 7⟌7 0

E 7⟌4 2

5

A   3
7 3⟌2 1 5

B   5
5 4⟌3 3 0

Part 5 continues on the next page.

C   8
46|419   [ ]

D   7
37|255   [ ]

E   3
37|94   [ ]

F   4
43|217   [ ]

6

A
57|752

B
22|946

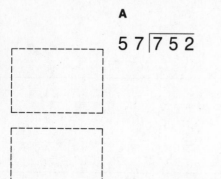

C
39|2528

D
70|1726

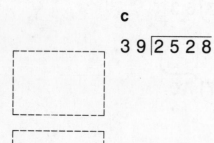

7

7|35   8|56   3|3   6|12   4|36   3|9   8|72

9|45   4|24   5|15   8|48   2|8   6|30   7|14

Part 7 continues on the next page.

$2\overline{)18}$ $9\overline{)36}$ $6\overline{)6}$ $7\overline{)21}$ $8\overline{)40}$ $4\overline{)28}$ $2\overline{)10}$

$8\overline{)64}$ $7\overline{)28}$ $9\overline{)27}$ $3\overline{)6}$ $6\overline{)24}$ $3\overline{)12}$ $4\overline{)40}$

$9\overline{)18}$ $6\overline{)18}$ $2\overline{)2}$ $4\overline{)32}$ $5\overline{)30}$ $2\overline{)16}$ $5\overline{)50}$

$5\overline{)35}$ $3\overline{)15}$ $5\overline{)20}$ $7\overline{)7}$ $5\overline{)25}$

8

$7\overline{)56}$ $7\overline{)42}$ $4\overline{)8}$ $7\overline{)70}$ $7\overline{)56}$ $4\overline{)12}$ $7\overline{)49}$

$7\overline{)63}$ $7\overline{)70}$ $7\overline{)42}$ $4\overline{)8}$ $7\overline{)63}$ $7\overline{)56}$ $7\overline{)70}$

$7\overline{)49}$ $4\overline{)12}$ $7\overline{)63}$ $4\overline{)12}$ $4\overline{)8}$

9

A $6\overline{)1949}$ B $8\overline{)5675}$ C $2\overline{)1119}$ D $6\overline{)9084}$ E $7\overline{)2135}$

10

Write the facts with no remainders.

A $6\overline{)9}$

B $6\overline{)32}$

C $6\overline{)20}$

D $6\overline{)3}$

E $6\overline{)17}$

F $6\overline{)26}$

1

A 4⟌1 2

B 4⟌1 6

C 4⟌8

D 4⟌2 0

2

A 4⟌4

B 4⟌8

C 4⟌1 2

D 4⟌1 6

E 4⟌2 0

3

A 3⟌1 2 → 4

B 2⟌1 2 → 6

4

A 7⟌6 3

B 7⟌4 9

C 7⟌4 2

D 7⟌7 0

E 7⟌5 6

5

A 2 2⟌7 9 4

B 3 5⟌3 2 4 3

C 5 6⟌2 3 2 7

Part 5 continues on the next page.

D
$$5\,1\,\overline{)2\,4\,8\,1}$$

E
$$7\,2\,\overline{)3\,8\,2\,9}$$

6

$$6\overline{)18} \quad 9\overline{)36} \quad 2\overline{)4} \quad 8\overline{)40} \quad 5\overline{)10} \quad 4\overline{)24} \quad 5\overline{)25}$$

$$10\overline{)90} \quad 5\overline{)40} \quad 6\overline{)6} \quad 3\overline{)9} \quad 9\overline{)18} \quad 4\overline{)40} \quad 10\overline{)30}$$

$$9\overline{)9} \quad 4\overline{)32} \quad 10\overline{)10} \quad 6\overline{)30} \quad 8\overline{)64} \quad 5\overline{)15} \quad 2\overline{)12}$$

$$9\overline{)27} \quad 7\overline{)14} \quad 3\overline{)6} \quad 8\overline{)72} \quad 6\overline{)12} \quad 7\overline{)28} \quad 3\overline{)12}$$

$$6\overline{)24} \quad 2\overline{)12} \quad 8\overline{)48} \quad 2\overline{)8} \quad 9\overline{)45} \quad 4\overline{)28} \quad 5\overline{)40}$$

$$3\overline{)15} \quad 8\overline{)56} \quad 7\overline{)35} \quad 4\overline{)36} \quad 7\overline{)21}$$

7

$$7\overline{)42} \quad 4\overline{)12} \quad 7\overline{)56} \quad 7\overline{)63} \quad 4\overline{)8} \quad 7\overline{)49} \quad 4\overline{)12}$$

$$7\overline{)63} \quad 7\overline{)56} \quad 4\overline{)8} \quad 7\overline{)42} \quad 4\overline{)12} \quad 7\overline{)70} \quad 7\overline{)56}$$

$$7\overline{)70} \quad 4\overline{)8} \quad 7\overline{)49} \quad 7\overline{)63} \quad 7\overline{)70}$$

8

Change the answer if the remainder is too big or too small.

A
$$\overset{3}{27\overline{)114}}$$

B
$$\overset{8}{56\overline{)473}}$$

C
$$\overset{4}{74\overline{)370}}$$

D
$$\overset{5}{63\overline{)307}}$$

E
$$\overset{7}{48\overline{)392}}$$

9

Write the facts with no remainders.

A

$$6\overline{)22}$$

B
$$6\overline{)15}$$

C
$$6\overline{)27}$$

D
$$6\overline{)10}$$

E
$$6\overline{)33}$$

F
$$6\overline{)25}$$

10

A
$$4\overline{)3584}$$

B
$$9\overline{)933}$$

C
$$6\overline{)1381}$$

D
$$3\overline{)1230}$$

E
$$5\overline{)41,879}$$

11

A Each box will hold 8 statues. There are 120 boxes. How many statues can be put in the boxes?

B There are 5 offices on each floor. There are 120 floors. How many offices are there?

C Art's CD company wants to make 2000 copies of a CD. His machine makes 5 CDs every minute. How many minutes will it take to make the CDs?

D Rose's company makes 525 sewing machines every day. They made sewing machines for 5 days. How many sewing machines did they make?

Lesson 47

Test + Facts + Problems + Bonus = TOTAL

1

A 6⟌5 4   B 6⟌3 6   C 6⟌4 8   D 6⟌4 2

2

A 6⟌3 6   B 6⟌4 2

C 6⟌4 8   D 6⟌5 4

E 6⟌6 0

3

A   B   C   D

4

A 4⟌2 0   B 4⟌4

C 4⟌1 6   D 4⟌8

E 4⟌1 2

5

A 5 3⟌4 0 1 4   B 6 5⟌5 3 2 8

Part 5 continues on the next page.

122 Lesson 47

C
24|607

D
62|1984

6

A 21|1542   B 27|1234   C 43|1824   D 37|1268

7

4|24   9|27   6|18   4|36   7|70   6|12   5|10

10|50   7|49   4|40   8|72   6|30   9|36   3|6

7|35   8|40   5|30   7|14   9|45   4|32   7|56

7|21   2|6   9|18   3|15   10|40   5|20   7|42

8|56   3|12   7|63   2|18   6|24   4|28   8|48

3|9   2|14   5|35   7|28   8|64

8

4) 1 6   6) 3 6   6) 4 2   4) 1 6   4) 2 0   4) 1 2   4) 1 6

6) 3 6   4) 1 2   4) 8   6) 4 2   4) 2 0   4) 8   6) 3 6

9

Change the answer if the remainder is too big or too small.

A
```
        3
6 6 ) 1 9 4
```

B
```
        5
4 7 ) 2 1 9
```

C
```
        2
5 3 ) 1 3 4
```

D
```
        7
8 4 ) 6 7 5
```

10

A A race horse ran 312 kilometers each week. It ran for 3 weeks. How many kilometers did it run?

B 424 people want to take a train ride. Each seat on the train holds 4 people. How many seats are needed?

C Angel, the car dealer, had 1920 cars. He put the same number of cars into each of 6 lots. How many cars did he put in each lot?

D It took Yoko 7 months to build a dam. 231 loads of rock were used each month. How many loads of rock went into the dam?

1

A 6⟌4 8   B 6⟌6 0   C 6⟌4 2   D 6⟌5 4

2

A 6⟌3 6

B 6⟌4 2

C 6⟌4 8

D 6⟌5 4

E 6⟌6 0

3

A 8 people got on an empty bus. At the first bus stop, 1 person got off. How many people end up on the bus?

[ ] people

B The store has 27 cakes. A truck brings 35 cakes. How many cakes does the store end up with?

[ ] cakes

C Kip has 6 books. He gets 12. How many books does Kip end up with?

[ ] books

D Sid has 7 cans of paint. He uses 6. How many cans does Sid end up with?

[ ] cans

E Rob has 9 pens. He got 4. How many pens does Rob have now?

[ ] pens

F Leon had 40 nails. He used 5 nails. How many nails does Leon have now?

[ ] nails

Part 3 continues on the next page.

G A farmer had 135 beets. He eats 14. How many beets does the farmer end up with?

 beets

4

A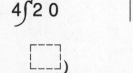
4⟌8

B 4⟌1 2

C 4⟌2 0

D 4⟌4

E 4⟌1 6

5

4⟌2 8	7⟌2 1	6⟌1 8	9⟌1 8	7⟌4 9	8⟌4 8	9⟌4 5
5⟌1 0	8⟌4 0	4⟌2 4	2⟌6	6⟌2 4	7⟌1 4	8⟌5 6
3⟌1 2	7⟌7	8⟌7 2	4⟌3 2	7⟌6 3	6⟌1 2	2⟌8
7⟌4 2	3⟌9	6⟌6	8⟌6 4	4⟌3 6	5⟌2 5	7⟌3 5
5⟌1 5	2⟌4	3⟌6	7⟌5 6	5⟌4 0	4⟌4 0	6⟌3 0
7⟌2 8	2⟌1 0	9⟌2 7	3⟌1 5	9⟌3 6		

6

6⟌3 6	4⟌1 6	6⟌4 2	4⟌8	4⟌2 0	4⟌1 6	6⟌4 2
4⟌2 0	6⟌3 6	4⟌1 2	4⟌2 0	6⟌4 2	4⟌8	4⟌1 6

7

Write the facts with no remainders.

A 7⟌7 1 ⌐

B 7⟌4 6 ⌐

C 7⟌6 5 ⌐

D 7⟌5 4 ⌐

E 7⟌6 0 ⌐

F 7⟌5 5 ⌐

8

A 5 4⟌4 5 4 2

B 7 8⟌5 2 3 4

C 4 2⟌2 7 1 6

9

A 9 4⟌3 0 6 5

B 7⟌2 9 4 2

C 3 9⟌2 9 7 2

D 8⟌8 1 6 8

10

Change the answer if the remainder is too big or too small.

A
   5
2 6⟌1 2 0 ⌐

B
   4
3 7⟌1 5 9 ⌐

C
   3
1 4⟌3 8 ⌐

Test + Facts + Problems + Bonus = TOTAL

1

A 6√4 2

B 6√3 6

C 6√5 4

D 6√6 0

E 6√4 8

2

A   B   C   D

3

A A man has 25 tools. 2 are picks and the rest are saws. How many saws does he have?

[ ] saws

B Greg has trees. 2 are pine trees and 38 are oak trees. How many trees does Greg have?

[ ] trees

C Clara has tools. She has 34 drills and 9 hammers. How many tools does she have?

[ ] tools

D Ryan has 9 flowers. He has 5 roses and the rest are tulips. How many tulips does Ryan have?

[ ] tulips

E Andy has 219 buttons. He has 106 blue buttons and the rest are black. How many black buttons does Andy have?

[ ] buttons

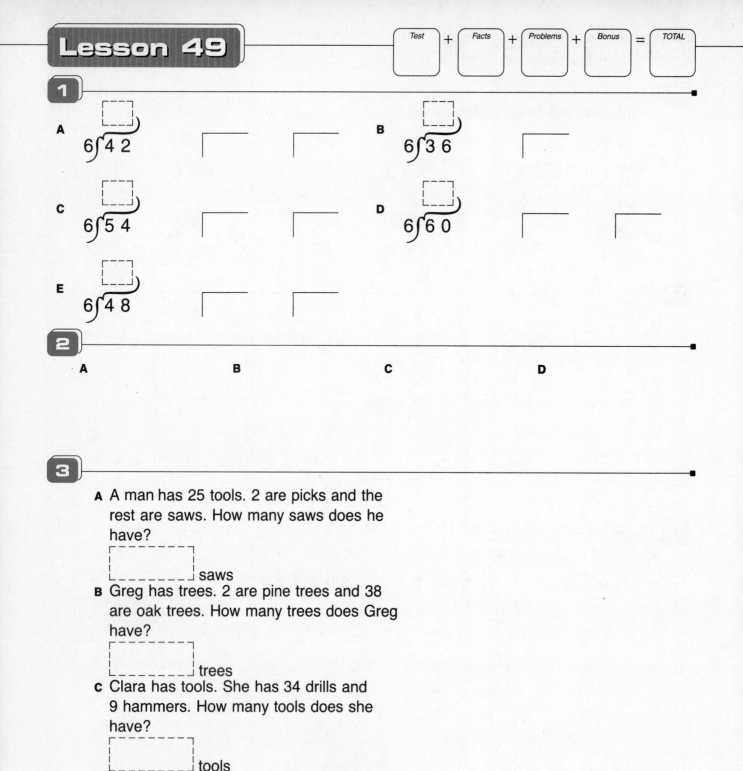

Part 3 continues on the next page.

F The farmer has animals. She has 40 pigs and 30 chickens. How many animals does the farmer have?

[____] animals

G Ms. Dodge's store has 54 chairs. There are 19 kitchen chairs and the rest are living room chairs. How many living room chairs does Ms. Dodge's store have?

[____] chairs

4

8)64	7)42	5)35	3)12	4)24	2)18	8)56
7)28	6)24	2)14	4)12	2)16	7)21	4)16
10)100	8)72	7)49	9)36	7)14	6)18	8)48
4)32	9)27	7)63	8)40	4)20	7)35	10)80
5)45	4)8	3)15	6)12	10)20	2)12	4)28
3)9	6)30	7)56	4)36	5)20		

5

6)42	6)60	6)54	6)36	4)16	6)48	6)54
4)20	6)60	6)36	6)54	6)48	6)42	6)60

6

Write the facts with no remainders.

A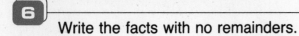

7 ⟌ 4 7

B

7 ⟌ 5 0

C

7 ⟌ 6 1

D

7 ⟌ 5 7

E

7 ⟌ 4 5

F

7 ⟌ 7 3

7

A   7 ⟌ 9 1 4

B   5 9 ⟌ 2 4 9 4

C   3 ⟌ 7 2 1 5

D   8 7 ⟌ 2 8 4 7

8

A   9 4 ⟌ 2 2 3 5

B   3 6 ⟌ 2 5 2

C   4 6 ⟌ 1 4 4 8

D   2 1 ⟌ 8 1 7

Lesson 50

1

A 6)5 4

B 6)6 0

C 6)3 6

D 6)4 8

E 6)4 2

2

A 7 4)7 9 3 4

B 5 8)1 2,0 7 5

C 2 1)8 5 0 1

D 6 3)3 8,3 8 7

3

A A kid has 37 toothpicks. His friend has 110 toothpicks. How many more toothpicks does his friend have than he does?

B Willie's turkey weighs 30 pounds. Alma's turkey weighs 25 pounds. How many pounds lighter is Alma's turkey than Willie's?

Part 3 continues on the next page.

c Hole A is 22 feet deep. Hole B is 4 feet deeper than hole A. How many feet deep is hole B?

┌ ─ ─ ─ ─ ─ ┐
                      ─ ─ ─ ─ ─ ─ ─ ─ ─ ─ ─
└ ─ ─ ─ ─ ─ ┘

d Herb has 139 seeds in a bag. His mom has 657 more seeds than he does. How many seeds does his mom have?

┌ ─ ─ ─ ─ ─ ┐
                      ─ ─ ─ ─ ─ ─ ─ ─ ─ ─ ─
└ ─ ─ ─ ─ ─ ┘

4

A
The first story problem that Soo works deals with the same number again and again.
+ −
× ÷

B
The next problem that Soo works does not deal with the same number again and again.
+ −
× ÷

C
The next problem that Soo works does not deal with the same number again and again.
+ −
× ÷

D
The next problem that Soo works deals with the same number again and again.
+ −
× ÷

E
The next problem that Soo works does not deal with the same number again and again.
+ −
× ÷

5

A

$58\overline{)4293}$

B

$24\overline{)789}$

C

$67\overline{)5034}$

D

$35\overline{)1524}$

6

7)14	8)56	4)12	9)36	9)27	4)36	5)25
6)18	7)63	5)40	8)72	3)6	4)20	2)10
5)50	4)8	7)21	2)4	7)35	5)10	9)18
4)28	7)56	6)12	8)40	6)30	2)20	4)24
3)9	7)49	2)16	4)32	8)64	7)28	3)12
7)42	4)16	8)48	9)45	6)24		

7

6)48	6)60	6)54	4)16	6)42	6)48	6)60
6)54	4)8	6)36	6)60	4)12	6)54	6)42

8

Write the facts with no remainders.

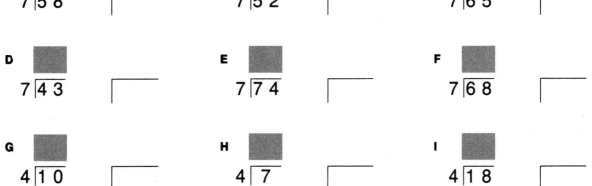

A. 7)58

B. 7)52

C. 7)65

D. 7)43

E. 7)74

F. 7)68

G. 4)10

H. 4)7

I. 4)18

Part 8 continues on the next page.

J  4)2 3

K 4)1 4

L 4)2

9

A 4)9 2 1 6

B 7 4)3 5 8 8

C 7)3 6 4 2

D 7 1)9 4 6

E 4)3 5 8 4

F 9)9 3 3

G 6)1 3 8 1

H 3)1 2 3 0

Lesson 51

Facts + Problems + Bonus = TOTAL

1

A 9⟌6 3

B 9⟌8 1

C 9⟌7 2

D 9⟌9 0

2

A 9⟌5 4

B 9⟌6 3

C 9⟌7 2

D 9⟌8 1

E 9⟌9 0

3

A   B   C   D

4

A Donna's store has 37 pairs of shoes. Mable's store has 110. How many more pairs of shoes does Mable's store have?

B There were 25 ships in the harbor on Monday. There were 30 ships in the harbor on Tuesday. How many more ships were in the harbor on Tuesday than on Monday?

C There are 22 cars at stoplight A. There are 4 more cars at stoplight B than at A. How many cars are at stoplight B?

D Curt's oil well is 139 meters deep. He drills 657 more meters. How many meters deep is Curt's oil well?

Lesson 51 ——•**135**

5

A

7 4 | 1 5,1 2 5

B

6 2 | 3 7,7 8 1

C

6 9 | 5 5,6 8 4

D

2 2 | 6 7 1 4

6

A

The first story problem that Sal works does not deal with the same number again and again.

+ −
× ÷

B

The next problem that Sal works deals with the same number again and again.

+ −
× ÷

C

The next problem that Sal works deals with the same number again and again.

+ −
× ÷

7

A

The problem does not deal with the same number again and again.
The problem does not give the big number.

+ −
× ÷

B

The problem deals with the same number again and again. The problem does not give the big number.

+ −
× ÷

C

The problem deals with the same number again and again. The problem gives the big number.

+ −
× ÷

D

The problem does not deal with the same number again and again.
The problem gives the big number.

+ −
× ÷

Part 7 continues on the next page.

E

The problem deals with the same number again and again. The problem does not give the big number.

+ −

× ÷

F

The problem deals with the same number again and again. The problem gives the big number.

+ −

× ÷

G

The problem does not deal with the same number again and again. The problem does not give the big number.

+ −

× ÷

H

The problem deals with the same number again and again. The problem does not give the big number.

+ −

× ÷

I

The problem does not deal with the same number again and again. The problem gives the big number.

+ −

× ÷

J

The problem deals with the same number again and again. The problem gives the big number.

+ −

× ÷

8

4)24	9)45	6)48	7)42	3)9	6)24	7)49
8)64	6)36	4)12	8)48	9)36	7)14	6)42
4)28	7)28	4)36	2)16	6)18	5)15	4)8
8)72	7)35	9)18	6)12	7)56	4)16	5)45
9)27	6)54	4)32	7)63	6)30	8)56	3)12
4)20	3)15	5)30	2)18	7)21		

9

9)63	4)16	9)54	9)72	9)63	6)42	9)54
6)54	6)36	4)12	9)54	4)20	9)63	6)48

10

Write the facts with no remainders.

A
4⟌2 1

B
4⟌1 1

C
4⟌1 9

D
4⟌6

E
4⟌1 5

F
4⟌3

11

A
1 5⟌7 0 5

B
6⟌7 3 2 4

C
2 6⟌1 3 4 8

D
8⟌8 4 0 8

E
2 1⟌1 5 4 2

F
2 7⟌1 2 3 4

G
4 3⟌1 8 2 4

H
3 7⟌1 2 6 8

Facts + Problems + Bonus = TOTAL

1

A 9⟌8 1

B 9⟌6 3

C 9⟌5 4

D 9⟌7 2

2

A 9⟌5 4

B 9⟌6 3

C 9⟌7 2

D 9⟌8 1

E 9⟌9 0

3

A

6⟌4 5

        7
6 4⟌4 4 9 6
  −4 4 8
        1

B

7 1⟌3 5 9 8

C

4 3⟌9 8 7

D

5 2⟌5 4 1 5

4

A

The problem deals with the same number again and again. The problem does not give the big number.

+ −

× ÷

B

The problem does not deal with the same number again and again. The problem does not give the big number.

+ −

× ÷

Part 4 continues on the next page.

C

The problem deals with the same number again and again. The problem gives the big number.

$+ \; -$
$\times \; \div$

D

The problem deals with the same number again and again. The problem does not give the big number.

$+ \; -$
$\times \; \div$

5

A There are 6 kids on each team. There are 48 kids. How many teams are there?

$+ \; - \; \times \; \div$

B Sherry has some leaves. 8 of them are red and 9 of them are yellow. How many leaves does Sherry have?

$+ \; - \; \times \; \div$

C The parade has 17 bands. 6 of the bands are on floats and the rest of them are marching. How many bands are marching?

$+ \; - \; \times \; \div$

D Pete did 35 problems on each page. He did 11 pages of problems. How many problems did he do?

$+ \; - \; \times \; \div$

E Betty types 7 pages an hour. She types 56 pages. How many hours did she type?

$+ \; - \; \times \; \div$

6

$4\overline{)36} \qquad 8\overline{)64} \qquad 7\overline{)21} \qquad 2\overline{)4} \qquad 7\overline{)56} \qquad 3\overline{)6} \qquad 7\overline{)35}$

$6\overline{)48} \qquad 5\overline{)35} \qquad 9\overline{)36} \qquad 6\overline{)18} \qquad 7\overline{)28} \qquad 8\overline{)48} \qquad 4\overline{)8}$

Part 6 continues on the next page.

7⟌7   4⟌2 0   5⟌2 0   8⟌7 2   1 0⟌3 0   6⟌2 4   3⟌9

7⟌4 2   4⟌2 4   8⟌4 0   6⟌4 2   1 0⟌6 0   9⟌2 7   4⟌2 8

9⟌4 5   6⟌5 4   7⟌4 9   3⟌1 2   7⟌6 3   5⟌4 0   4⟌1 2

4⟌1 6   2⟌1 4   8⟌5 6   2⟌1 2   4⟌3 6

7

9⟌6 3   6⟌4 2   9⟌5 4   6⟌6 0   9⟌7 2   9⟌6 3   9⟌4 5

9⟌7 2   9⟌5 4   4⟌8   6⟌3 6   9⟌6 3   6⟌4 8   9⟌5 4

8

Write the facts with no remainders.

A ▢ 4⟌5   ⌐

B ▢ 4⟌2 2   ⌐

C ▢ 4⟌1 3   ⌐

D ▢ 4⟌9   ⌐

E ▢ 4⟌1   ⌐

F ▢ 4⟌1 7   ⌐

G ▢ 6⟌4 6   ⌐

H ▢ 6⟌3 8   ⌐

I ▢ 6⟌6 4   ⌐

J ▢ 6⟌5 0   ⌐

K ▢ 6⟌5 5   ⌐

L ▢ 6⟌4 3   ⌐

9

A 3 8⟌2 7 3 9   B 4⟌4 1 3 5   C 2 4⟌6 0 7   D 4 7⟌1 4,3 8 6   E 7⟌9 1 5

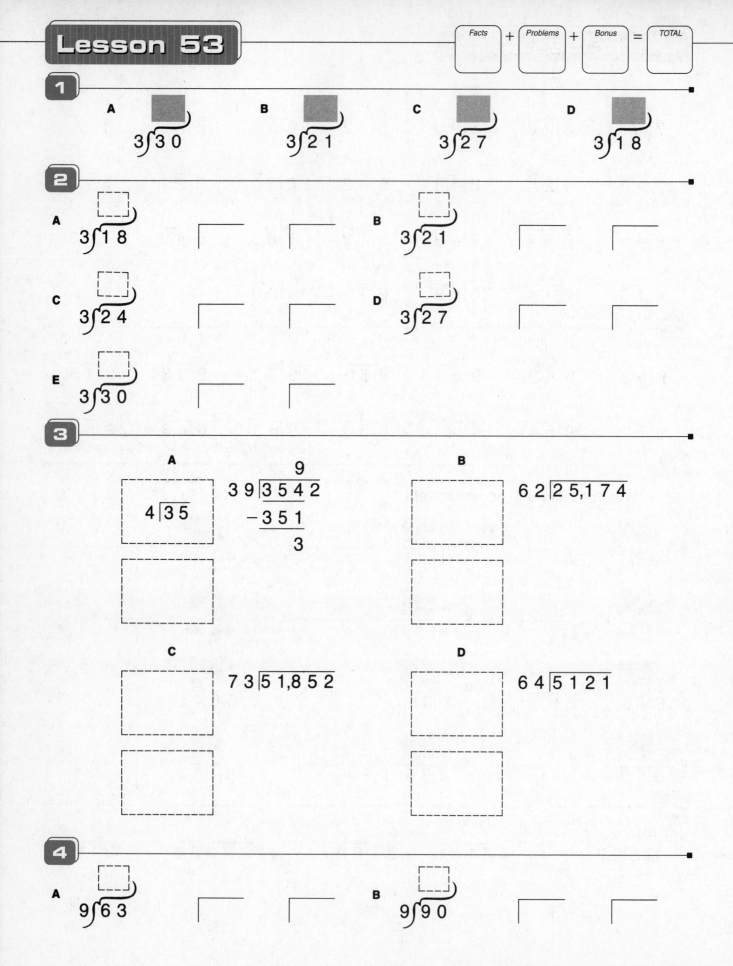

Facts + Problems + Bonus = TOTAL

1

A 3⟌3 0   B 3⟌2 1   C 3⟌2 7   D 3⟌1 8

2

A 3⟌1 8   B 3⟌2 1

C 3⟌2 4   D 3⟌2 7

E 3⟌3 0

3

A
4⟌3 5

```
       9
39⟌3 5 4 2
  −3 5 1
        3
```

B
6 2⟌2 5,1 7 4

C
7 3⟌5 1,8 5 2

D
6 4⟌5 1 2 1

4

A 9⟌6 3   B 9⟌9 0

Part 4 continues on the next page.

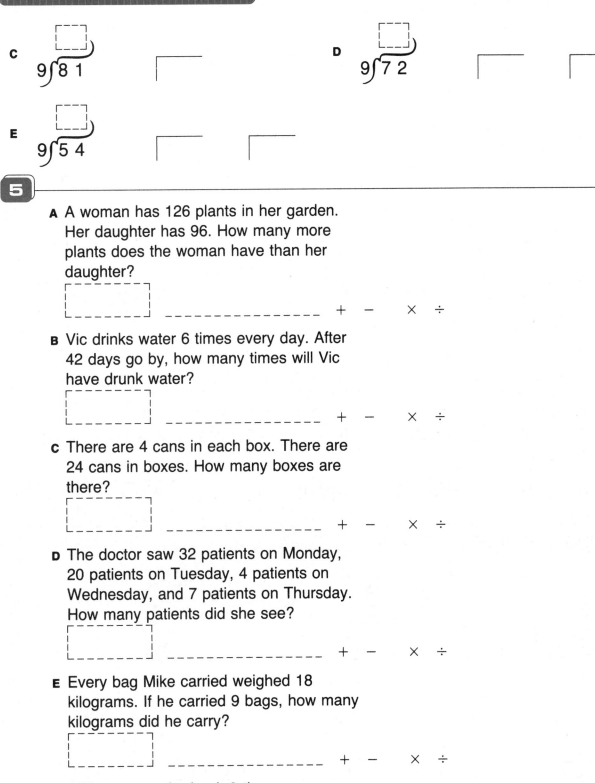

C 9 ⌐ 8 1

D 9 ⌐ 7 2

E 9 ⌐ 5 4

5

A A woman has 126 plants in her garden. Her daughter has 96. How many more plants does the woman have than her daughter?

+ − × ÷

B Vic drinks water 6 times every day. After 42 days go by, how many times will Vic have drunk water?

+ − × ÷

C There are 4 cans in each box. There are 24 cans in boxes. How many boxes are there?

+ − × ÷

D The doctor saw 32 patients on Monday, 20 patients on Tuesday, 4 patients on Wednesday, and 7 patients on Thursday. How many patients did she see?

+ − × ÷

E Every bag Mike carried weighed 18 kilograms. If he carried 9 bags, how many kilograms did he carry?

+ − × ÷

F Millie went to the bank 9 times every month. In all, she went to the bank 126 times. How many months did Millie go to the bank?

+ − × ÷

6

8⟌72	7⟌42	3⟌12	7⟌21	4⟌28	9⟌36	5⟌30
5⟌25	1⟌6	7⟌49	2⟌6	6⟌18	7⟌28	8⟌56
1⟌2	7⟌56	9⟌9	4⟌0	8⟌48	4⟌20	2⟌10
6⟌30	3⟌6	4⟌12	5⟌10	9⟌27	6⟌6	7⟌0
7⟌63	6⟌42	2⟌8	8⟌64	10⟌40	7⟌35	3⟌15
6⟌54	4⟌36	5⟌45	10⟌10	5⟌5		

7

9⟌90	9⟌63	9⟌81	9⟌54	3⟌18	9⟌90	3⟌21
3⟌18	9⟌81	3⟌21	9⟌63	3⟌18	9⟌54	9⟌81

8

Write the facts with no remainders.

A 6⟌39

B 6⟌58

C 6⟌62

D 6⟌44

E 6⟌40

F 6⟌49

9

A 35⟌2843   B 6⟌2523   C 20⟌310   D 48⟌240   E 62⟌1814

Lesson 54

1

A
3√2 4

B
3√1 8

C
3√3 0

D
3√2 1

2

A
3√1 8

B
3√2 1

C
3√2 4

D
3√2 7

E
3√3 0

3

A
8 4√8 2 5

B
3 3√3 0 9

4

A
9√7 2

B
9√5 4

C
9√9 0

D
9√8 1

E
9√6 3

5

A	B	C	D	E
6$\overline{)3020}$	75$\overline{)3052}$	34$\overline{)3618}$	43$\overline{)2186}$	15$\overline{)452}$

6

2$\overline{)16}$ 7$\overline{)28}$ 4$\overline{)32}$ 8$\overline{)40}$ 6$\overline{)48}$ 5$\overline{)35}$ 4$\overline{)16}$

1$\overline{)8}$ 9$\overline{)18}$ 10$\overline{)50}$ 4$\overline{)24}$ 7$\overline{)14}$ 8$\overline{)64}$ 6$\overline{)24}$

6$\overline{)60}$ 7$\overline{)35}$ 3$\overline{)9}$ 2$\overline{)2}$ 8$\overline{)72}$ 4$\overline{)8}$ 6$\overline{)12}$

4$\overline{)4}$ 8$\overline{)48}$ 2$\overline{)18}$ 4$\overline{)40}$ 9$\overline{)45}$ 6$\overline{)36}$ 3$\overline{)3}$

5$\overline{)40}$ 10$\overline{)70}$ 8$\overline{)80}$ 5$\overline{)15}$ 1$\overline{)4}$ 8$\overline{)56}$ 5$\overline{)50}$

10$\overline{)90}$ 2$\overline{)20}$ 7$\overline{)49}$ 5$\overline{)20}$ 7$\overline{)56}$

7

9$\overline{)72}$ 3$\overline{)21}$ 9$\overline{)81}$ 3$\overline{)18}$ 9$\overline{)54}$ 9$\overline{)90}$ 3$\overline{)21}$

9$\overline{)90}$ 9$\overline{)63}$ 3$\overline{)18}$ 9$\overline{)72}$ 3$\overline{)21}$ 9$\overline{)81}$ 9$\overline{)54}$

9$\overline{)63}$ 9$\overline{)90}$ 9$\overline{)81}$ 3$\overline{)21}$

8

Write the facts with no remainders.

A 6⟌5 2

B 6⟌4 1

C 6⟌4 7

D 6⟌3 7

E 6⟌5 3

F 6⟌5 6

9

Don't forget to circle the right sign.

A Hugo wants to make 42 stacks of boxes. If he puts 126 boxes in every stack, how many boxes will he use?

+ − × ÷

B There are 3 eggs in each nest. There are 99 eggs in nests. How many nests are there?

+ − × ÷

C Bessie's report has 14 pages, Del's report has 22 pages, and Liz's report has 47 pages. How many pages are there in all?

+ − × ÷

D Airplanes land at Midtown Airport for 6 days. 41 airplanes land each day. How many airplanes land?

+ − × ÷

Facts $+$ Problems $+$ Bonus $=$ TOTAL

1

A 8⟌3 2 B 8⟌1 6 C 8⟌8 D 8⟌2 4

2

A 8⟌8

B 8⟌1 6

C 8⟌2 4

D 8⟌3 2

3

A   B   C   D

4

A 3⟌2 7

B 3⟌3 0

C 3⟌2 4

D 3⟌1 8

E 3⟌2 1

5

A 4 3⟌4 2 8   B 8 4⟌8 1 5   C 6 2⟌6 1 2

6

4⟌2 0	6⟌4 2	1 0⟌8 0	9⟌8 1	9⟌6 3	2⟌1 4	5⟌0
6⟌6	7⟌6 3	9⟌2 7	8⟌4 8	4⟌1 2	1 0⟌2 0	3⟌1 5
8⟌6 4	9⟌7 2	4⟌0	3⟌6	4⟌2 8	9⟌3 6	2⟌1 2
5⟌3 0	7⟌7 0	6⟌1 8	4⟌3 6	8⟌5 6	5⟌4 5	9⟌9 0
9⟌5 4	8⟌7 2	7⟌2 1	3⟌1 2	1⟌7	2⟌4	1 0⟌0
5⟌2 5	9⟌0	6⟌5 4	7⟌4 2	6⟌3 0		

7

3⟌2 1	3⟌3 0	3⟌2 4	8⟌1 6	3⟌1 8	8⟌2 4	3⟌2 4
8⟌1 6	3⟌1 8	3⟌3 0	3⟌2 7	8⟌2 4	8⟌1 6	3⟌3 0
3⟌2 7	3⟌2 4	8⟌2 4	3⟌2 1	3⟌2 7		

8

Don't forget to circle the right sign.

A Mack has 28 boards. 19 of the boards are
long. The rest of the boards are short.
How many short boards does Mack have?

```
┌ ─ ─ ─ ─ ┐
│         │    ─ ─ ─ ─ ─ ─ ─ ─ ─    +  ─    ×  ÷
└ ─ ─ ─ ─ ┘
```

Part 8 continues on the next page.

B Jan rides her bike to school 3 days a week. She rides it to school 141 times. How many weeks did Jan ride her bike to school?

+ − × ÷

C Every time Sara goes to the creek, she gets 8 rocks. She goes to the creek 34 times. How many rocks does she get?

+ − × ÷

D Lisa had 347 nuts. Then she found some more. Now she has 872. How many did she find?

+ − × ÷

9

A 22⟌6714

B 58⟌2341

C 26⟌784

D 7⟌8423

E 24⟌789

F 8⟌1000

Facts + Problems + Bonus = TOTAL

1

A 8⟌1 6

B 8⟌3 2

C 8⟌2 4

D 8⟌8

2

A 8⟌8

B 8⟌1 6

C 8⟌2 4

D 8⟌3 2

3

A
7 4⟌7 5 8

B
8 2⟌8 1 6

C
5 3⟌5 2 8

4

A 3⟌2 4

B 3⟌1 8

C 3⟌3 0

D 3⟌2 1

E 3⟌2 7

5

9)45	8)56	4)24	7)28	5)40	7)14	9)54
8)64	9)0	7)7	2)6	9)90	8)40	3)3
7)35	8)48	4)16	3)9	10)30	9)18	2)8
4)8	6)0	9)63	2)10	7)49	6)12	3)0
2)0	10)100	9)81	5)10	6)48	9)72	4)40
8)72	6)36	4)32	7)56	6)24		

6

3)21	3)30	3)24	8)16	3)27	8)24	3)30
3)27	3)24	3)18	8)24	3)30	3)21	8)16
3)18	3)27	3)24	8)16	8)24		

7

Write the facts with no remainders.

A
9)57

B
9)76

C
9)83

D
9)68

E
9)59

F
9)87

8

A Gus made dinner 3 times each week. He made dinner for 28 weeks. How many times did he make dinner?

```
┌──────────┐
│          │
└──────────┘  ─ ─ ─ ─ ─ ─ ─ ─ ─ ─ ─ ─ ─
```

B There are 24 girls at Friendly School. There are 21 boys and 6 teachers. How many people are at Friendly School?

```
┌────────┐
│        │
└────────┘  ─ ─ ─ ─ ─ ─ ─ ─ ─ ─ ─ ─
```

C Mr. Bear hooked up 62 telephones every month. He hooked up 434 telephones. How many months did Mr. Bear hook up telephones?

```
┌────────┐
│        │
└────────┘  ─ ─ ─ ─ ─ ─ ─ ─ ─ ─ ─ ─
```

D The town has 37 stores. 14 of them are closed on Saturdays. The rest are open. How many stores are open?

```
┌────────┐
│        │
└────────┘  ─ ─ ─ ─ ─ ─ ─ ─ ─ ─ ─ ─
```

E Every year Jackie got 41 letters. She got 369 letters. How many years went by?

```
┌────────┐
│        │
└────────┘  ─ ─ ─ ─ ─ ─ ─ ─ ─ ─ ─ ─
```

F Marjie chewed 8 pieces of celery every week. She chewed 64 pieces of celery. How many weeks did Marjie chew celery?

```
┌────────┐
│        │
└────────┘  ─ ─ ─ ─ ─ ─ ─ ─ ─ ─ ─ ─
```

G Patty dug 5 holes each hour. She dug holes for 230 hours. How many holes did Patty dig?

```
┌────────┐
│        │
└────────┘  ─ ─ ─ ─ ─ ─ ─ ─ ─ ─ ─ ─
```

9

A

$$33\overline{)2003}$$

B

$$65\overline{)2608}$$

C

$$80\overline{)3421}$$

D

$$8\overline{)5764}$$

E

$$35\overline{)1191}$$

F

$$24\overline{)1315}$$

G

$$50\overline{)3100}$$

H

$$52\overline{)5178}$$

I

$$37\overline{)958}$$

Lesson 57

1

A 8⟌2 4

B 8⟌1 6

C 8⟌3 2

D 8⟌8

2

A   B   C   D

3

| 3⟌2 1 | 2⟌1 8 | 7⟌4 2 | 9⟌8 1 | 4⟌2 8 | 8⟌5 6 | 1 0⟌6 0 |

| 6⟌3 0 | 8⟌7 2 | 2⟌1 6 | 3⟌2 7 | 7⟌2 1 | 9⟌3 6 | 4⟌3 6 |

| 2⟌4 | 3⟌2 4 | 9⟌2 7 | 4⟌1 2 | 3⟌6 | 7⟌0 | 3⟌3 0 |

| 9⟌9 | 3⟌1 2 | 8⟌4 8 | 7⟌7 0 | 2⟌2 0 | 9⟌6 3 | 4⟌2 0 |

| 6⟌1 8 | 2⟌2 | 9⟌7 2 | 7⟌6 3 | 6⟌4 2 | 5⟌2 0 | 9⟌9 0 |

| 6⟌5 4 | 3⟌1 8 | 9⟌5 4 | 8⟌6 4 | 3⟌1 5 |

4

| 3⟌1 8 | 8⟌3 2 | 8⟌1 6 | 3⟌2 4 | 8⟌3 2 | 9⟌6 3 | 3⟌2 7 |

| 9⟌8 1 | 8⟌1 6 | 3⟌2 1 | 8⟌3 2 | 9⟌5 4 | 8⟌2 4 | 9⟌7 2 |

5

A Jimmy has toys. Ten are yo-yos and 15 are trucks. How many toys does he have?

B Gussie bought oil 13 times every year. She bought oil 65 times. How many years did Gussie buy oil?

C Mr. Eagle used 3 eggs each time he baked bread. In all, he used 18 eggs to make bread. How many times did he bake bread?

D Every year Lucy went to the park 16 times. She went for 42 years. How many times did Lucy go to the park?

E Ming baby-sat 7 times each month. He baby-sat 168 times. How many months did Ming baby-sit?

F Marty tested 27 radios each day. He tested 189 radios. How many days did Marty test radios?

6

A 31|3 0 2 4   B 21|8 6 0 1   C 46|2 3 0 4   D 4|1 0 0 1

E 7|7 0 9 1   F 24|2 3 1 8   G 43|9 8 7

7

Write the facts with no remainders.

A 9|9 3

B 9|8 5

C 9|5 6

D 9|7 7

E 9|7 0

F 9|6 1

G 9|5 0

H 9|9 7

I 9|8 8

J 9|3 8

K 9|7 5

L 9|6 9

Facts + Problems + Bonus = TOTAL

1

A 8)2 4

B 8)8

C 8)3 2

D 8)1 6

2

9)7 2   3)1 8   6)3 6   7)3 5   8)7 2   9)4 5   4)2 8

4)3 2   9)6 3   7)5 6   8)6 4   3)9   2)1 4   6)4 8

3)2 1   8)5 6   4)4   9)1 8   4)2 4   7)7   8)8 0

4)1 6   9)8 1   8)4 0   7)2 8   1 0)4 0   3)3   2)1 2

6)6   8)4 8   3)1 5   6)2 4   7)1 4   5)5   3)2 4

3)2 7   3)1 2   9)5 4   7)4 9   5)3 5

3

8)3 2   9)5 4   3)3 0   8)1 6   3)2 1   8)3 2   9)7 2

9)6 3   8)3 2   3)1 8   9)8 1   8)2 4   3)2 7   8)1 6

4

A A swimmer swam 8 kilometers every hour. She swam 96 kilometers. How many hours did she swim?

Part 4 continues on the next page.

B The train made 3 stops every day it ran.
It made 72 stops. How many days did the
train run?

```
┌ ─ ─ ─ ─ ─ ┐
│           │    ─ ─ ─ ─ ─ ─ ─ ─ ─ ─ ─ ─ ─ ─
└ ─ ─ ─ ─ ─ ┘
```

C Maria has 155 pens. She keeps 42 pigs in
each pen. How many pigs does she have?

```
┌ ─ ─ ─ ─ ─ ┐
│           │    ─ ─ ─ ─ ─ ─ ─ ─ ─ ─ ─ ─ ─ ─
└ ─ ─ ─ ─ ─ ┘
```

D The mover lifted 23 boxes every hour.
He lifted 138 boxes. How many hours
did he lift boxes?

```
┌ ─ ─ ─ ─ ─ ┐
│           │    ─ ─ ─ ─ ─ ─ ─ ─ ─ ─ ─ ─ ─ ─
└ ─ ─ ─ ─ ─ ┘
```

E Carol wants to build 58 houses. She
needs to hire 12 people to work on each
house. How many people does she need
to hire?

```
┌ ─ ─ ─ ─ ─ ┐
│           │    ─ ─ ─ ─ ─ ─ ─ ─ ─ ─ ─ ─ ─ ─
└ ─ ─ ─ ─ ─ ┘
```

F Toni grew 15 carrots in 2002. She grew
11 carrots in 2004. How many more
carrots did she grow in 2002 than in 2004?

```
┌ ─ ─ ─ ─ ─ ┐
│           │    ─ ─ ─ ─ ─ ─ ─ ─ ─ ─ ─ ─ ─ ─
└ ─ ─ ─ ─ ─ ┘
```

5

A 42$\overline{)376}$ B 87$\overline{)9481}$ C 36$\overline{)2882}$ D 13$\overline{)1000}$

Part 5 continues on the next page.

E 99$\overline{)9909}$ F 54$\overline{)678}$ G 7$\overline{)2111}$

6

Write the facts with no remainders.

A 9$\overline{)74}$

B 9$\overline{)96}$

C 9$\overline{)60}$

D 9$\overline{)88}$

E 9$\overline{)67}$

F 9$\overline{)84}$

G 3$\overline{)31}$

H 3$\overline{)23}$

I 3$\overline{)20}$

J 3$\overline{)25}$

K 3$\overline{)22}$

L 3$\overline{)28}$

M 9$\overline{)57}$

N 9$\overline{)78}$

O 9$\overline{)64}$

Facts + Problems + Bonus = TOTAL

1

A   B   C   D

2

$8\overline{)3\,2}$ $7\overline{)4\,2}$ $10\overline{)7\,0}$ $9\overline{)6\,3}$ $8\overline{)8}$ $8\overline{)2\,4}$ $6\overline{)4\,8}$

$4\overline{)3\,2}$ $5\overline{)4\,5}$ $2\overline{)1\,8}$ $6\overline{)5\,4}$ $4\overline{)2\,8}$ $8\overline{)6\,4}$ $7\overline{)4\,9}$

$2\overline{)1\,6}$ $6\overline{)3\,6}$ $8\overline{)8}$ $3\overline{)2\,4}$ $7\overline{)5\,6}$ $9\overline{)8\,1}$ $5\overline{)3\,5}$

$8\overline{)5\,6}$ $5\overline{)4\,0}$ $4\overline{)2\,4}$ $9\overline{)5\,4}$ $3\overline{)2\,7}$ $8\overline{)1\,6}$ $7\overline{)6\,3}$

$10\overline{)6\,0}$ $7\overline{)3\,5}$ $2\overline{)1\,4}$ $8\overline{)4\,8}$ $3\overline{)1\,8}$ $6\overline{)4\,2}$ $5\overline{)3\,0}$

$3\overline{)2\,1}$ $9\overline{)7\,2}$ $8\overline{)4\,0}$ $4\overline{)3\,6}$ $8\overline{)7\,2}$

3

A $\quad 42\overline{)4\,0\,3\,5}$ B $\quad 31\overline{)6\,8\,4}$ C $\quad 7\overline{)9\,8\,1}$

D $\quad 9\overline{)8\,1\,0\,6}$ E $\quad 4\overline{)2\,8\,1\,8}$ F $\quad 67\overline{)5\,0\,3\,4}$

4

A Dennis drank 2 glasses of milk every time he ate lunch. He drank 24 glasses of milk at lunches. How many lunches did he have?

```
┌─ ─ ─ ─ ─ ─ ─ ┐
│             │   ─ ─ ─ ─ ─ ─ ─ ─ ─ ─ ─ ─ ─
└─ ─ ─ ─ ─ ─ ─ ┘
```

B Band A has 4 drum players, band B has 2 drum players, and band C has 11 drum players. How many drum players are there?

```
┌─ ─ ─ ─ ─ ─ ─ ┐
│             │   ─ ─ ─ ─ ─ ─ ─ ─ ─ ─ ─ ─ ─
└─ ─ ─ ─ ─ ─ ─ ┘
```

C Jenny milked 26 goats every night. She milked 78 goats. How many nights did Jenny milk goats?

```
┌─ ─ ─ ─ ─ ─ ─ ┐
│             │   ─ ─ ─ ─ ─ ─ ─ ─ ─ ─ ─ ─ ─
└─ ─ ─ ─ ─ ─ ─ ┘
```

D Faye grows 4 centimeters each year. After 8 years go by, how many centimeters will she have grown?

```
┌─ ─ ─ ─ ─ ─ ─ ┐
│             │   ─ ─ ─ ─ ─ ─ ─ ─ ─ ─ ─ ─ ─
└─ ─ ─ ─ ─ ─ ─ ┘
```

E A man must find 482 people to work for him. He finds 193 people the first day. How many more people must he find?

```
┌─ ─ ─ ─ ─ ─ ─ ┐
│             │   ─ ─ ─ ─ ─ ─ ─ ─ ─ ─ ─ ─ ─
└─ ─ ─ ─ ─ ─ ─ ┘
```

F Lonnie bought 3 liters of milk whenever he went to the store. He bought 21 liters of milk. How many times did he go to the store?

```
┌─ ─ ─ ─ ─ ─ ─ ┐
│             │   ─ ─ ─ ─ ─ ─ ─ ─ ─ ─ ─ ─ ─
└─ ─ ─ ─ ─ ─ ─ ┘
```

5

Write the facts with no remainders.

A 3)26

B 3)29

C 3)19

D 3)32

E 3)22

F 3)20

1

A Every square has 4 sides. There are 572 squares. How many sides are there?

┌ ─ ─ ─ ─ ─ ┐
└ ─ ─ ─ ─ ─ ┘ ─ ─ ─ ─ ─ ─ ─ ─ ─ ─ ─ ─ ─

B Angel has 163 friends. 44 are girls and the rest are boys. How many of Angel's friends are boys?

┌ ─ ─ ─ ─ ─ ┐
└ ─ ─ ─ ─ ─ ┘ ─ ─ ─ ─ ─ ─ ─ ─ ─ ─ ─ ─ ─

C Shannon took 2 breaks each day that she worked. She worked for 18 days. How many breaks did Shannon take?

┌ ─ ─ ─ ─ ─ ┐
└ ─ ─ ─ ─ ─ ┘ ─ ─ ─ ─ ─ ─ ─ ─ ─ ─ ─ ─ ─

D Rick has 24 pieces of pipe. He gets 17 pieces. How many pieces of pipe does Rick have now?

┌ ─ ─ ─ ─ ─ ┐
└ ─ ─ ─ ─ ─ ┘ ─ ─ ─ ─ ─ ─ ─ ─ ─ ─ ─ ─ ─

E Carmen painted 16 pictures every month. She painted 64 pictures. How many months did Carmen paint pictures?

┌ ─ ─ ─ ─ ─ ┐
└ ─ ─ ─ ─ ─ ┘ ─ ─ ─ ─ ─ ─ ─ ─ ─ ─ ─ ─ ─

F Gerald walked 9 blocks to work every day. He walked 90 blocks. How many days did he go to work?

┌ ─ ─ ─ ─ ─ ┐
└ ─ ─ ─ ─ ─ ┘ ─ ─ ─ ─ ─ ─ ─ ─ ─ ─ ─ ─ ─

2

$9\overline{)54}$ $7\overline{)49}$ $3\overline{)18}$ $4\overline{)16}$ $2\overline{)14}$ $8\overline{)24}$ $2\overline{)6}$

$5\overline{)25}$ $8\overline{)8}$ $7\overline{)63}$ $3\overline{)27}$ $9\overline{)36}$ $2\overline{)18}$ $8\overline{)32}$

Part 2 continues on the next page.

4⟌32   7⟌56   4⟌12   7⟌42   8⟌16   5⟌35   9⟌81

3⟌30   5⟌20   9⟌63   3⟌24   2⟌16   4⟌24   2⟌12

5⟌15   4⟌8   2⟌10   4⟌36   9⟌18   9⟌72   1⟌4

1⟌7   4⟌20   9⟌27   9⟌90   4⟌28

3

A   **B**   **C**
7⟌321   19⟌4017   4⟌581

D   **E**   **F**
6⟌1110   11⟌3789   81⟌9004

Part 3 continues on the next page.

G
$$65\overline{)2608}$$

H
$$35\overline{)1191}$$

I
$$80\overline{)3421}$$

4

Write the facts with no remainders.

A
$$3\overline{)25}$$

B
$$3\overline{)19}$$

C
$$3\overline{)31}$$

D
$$3\overline{)23}$$

E
$$3\overline{)32}$$

F
$$3\overline{)28}$$

G
$$8\overline{)4}$$

H
$$8\overline{)20}$$

I
$$8\overline{)36}$$

J
$$8\overline{)13}$$

K
$$8\overline{)34}$$

L
$$8\overline{)23}$$

1

6⟌3 6	9⟌6 3	4⟌1 6	2⟌1 6	3⟌2 1	6⟌1 2	2⟌1 4
5⟌3 5	7⟌5 6	6⟌5 4	9⟌7 2	7⟌2 1	4⟌2 0	6⟌4 2
6⟌2 4	3⟌2 4	8⟌4 8	6⟌1 8	1 0⟌4 0	9⟌5 4	7⟌4 9
6⟌4 8	3⟌9	3⟌2 7	7⟌6 3	7⟌4 2	8⟌1 6	4⟌8
3⟌1 8	4⟌4 0	9⟌8 1	2⟌6	3⟌3 0	1 0⟌1 0 0	9⟌1 8
8⟌3 2	4⟌1 2	8⟌2 4	6⟌3 0	5⟌4 5		

2

A June saw 17 birds in one tree and 14 birds in another tree. She saw 22 birds on the ground. How many birds did June see?

B Alice's company built 13 houses every week. They built 494 houses. How many weeks did they build houses?

C All triangles have 3 sides. Carlos drew 378 triangles. How many sides did he draw?

D There were 1140 sailboats sailing on lakes. There were 76 sailboats on every lake. How many lakes had sailboats?

Part 2 continues on the next page.

E All motorcycles have 2 wheels. There are 478 motorcycles. How many wheels are there?

```
┌ ─ ─ ─ ─ ┐
│         │     ─ ─ ─ ─ ─ ─ ─ ─ ─ ─ ─ ─
└ ─ ─ ─ ─ ┘
```

3

A	B	C	D	E
36⟌4992	17⟌9841	2⟌6819	9⟌648	8⟌9656

F	G	H	I	J
23⟌2264	34⟌3382	58⟌1357	3⟌918	26⟌3948

4

Write the facts with no remainders.

A ▨
  8⟌35   ┌

B ▨
  8⟌12   ┌

C ▨
  8⟌28   ┌

D ▨
  8⟌6   ┌

E ▨
  8⟌21   ┌

F ▨
  8⟌15   ┌

1

$9\overline{)72}$ $5\overline{)10}$ $4\overline{)36}$ $3\overline{)12}$ $8\overline{)16}$ $5\overline{)30}$ $4\overline{)16}$

$4\overline{)20}$ $9\overline{)81}$ $3\overline{)9}$ $6\overline{)24}$ $5\overline{)25}$ $7\overline{)63}$ $9\overline{)63}$

$3\overline{)27}$ $5\overline{)15}$ $8\overline{)24}$ $6\overline{)12}$ $7\overline{)42}$ $3\overline{)15}$ $5\overline{)35}$

$6\overline{)18}$ $3\overline{)6}$ $4\overline{)24}$ $9\overline{)54}$ $5\overline{)45}$ $6\overline{)30}$ $4\overline{)32}$

$8\overline{)32}$ $4\overline{)8}$ $3\overline{)24}$ $5\overline{)20}$ $8\overline{)8}$ $3\overline{)18}$ $5\overline{)40}$

$6\overline{)42}$ $4\overline{)28}$ $6\overline{)36}$ $3\overline{)21}$ $4\overline{)12}$

2

A Each bowl has 78 pieces of popcorn in it. If there are 2028 pieces of popcorn in bowls, how many bowls are there?

B There are 8 bottles in each case. There are 64 cases. How many bottles are there?

C I have 756 pieces of cake. Every cake has 9 pieces. How many cakes do I have?

D Amy drove 214 kilometers. Lola drove 108 kilometers. How many more kilometers did Amy drive than Lola?

Part 2 continues on the next page.

E Nathan has 476 coins. Roy has 58 more coins than Nathan. How many coins does Roy have?

```
┌─────────┐
│         │  ─ ─ ─ ─ ─ ─ ─ ─ ─ ─ ─
└─────────┘
```

F Heather buys 486 pieces of wood. She wants to make tables that use 81 pieces each. How many tables can she make?

```
┌─────────┐
│         │  ─ ─ ─ ─ ─ ─ ─ ─ ─ ─ ─
└─────────┘
```

G There are 24 girls at Clarksville School. There are 21 boys and 6 teachers. How many people are at Clarksville School?

```
┌─────────┐
│         │  ─ ─ ─ ─ ─ ─ ─ ─ ─ ─ ─
└─────────┘
```

H The shopping center has 37 stores. 14 of them are closed on Sundays. The rest are open. How many stores are open?

```
┌─────────┐
│         │  ─ ─ ─ ─ ─ ─ ─ ─ ─ ─ ─
└─────────┘
```

I Every year Nina went to the park 16 times. She went for 42 years. How many times did Nina go to the park?

```
┌─────────┐
│         │  ─ ─ ─ ─ ─ ─ ─ ─ ─ ─ ─
└─────────┘
```

3

A	B	C	D	E
8⟌987	58⟌4293	74⟌53,956	11⟌1385	9⟌780

Part 3 continues on the next page.

F $5\overline{)7993}$ G $74\overline{)7148}$ H $3\overline{)2869}$ I $89\overline{)8740}$ J $5\overline{)2345}$

4

Write the facts with no remainders.

A $8\overline{)19}$

B $8\overline{)10}$

C $8\overline{)38}$

D $8\overline{)26}$

E $8\overline{)17}$

F $8\overline{)31}$

Facts + Problems + Bonus = TOTAL

1

2)6̄	5)3̄0̄	4)2̄0̄	3)1̄2̄	5)1̄0̄	2)1̄2̄	7)4̄9̄
3)9̄	2)1̄8̄	5)1̄5̄	3)3̄	4)1̄2̄	6)4̄2̄	4)3̄2̄
4)1̄6̄	6)3̄6̄	3)6̄	2)1̄0̄	4)8̄	6)1̄8̄	3)2̄1̄
4)3̄6̄	2)8̄	3)1̄8̄	8)6̄4̄	4)2̄8̄	5)2̄5̄	7)5̄6̄
3)1̄5̄	9)7̄2̄	6)1̄2̄	4)2̄4̄	7)6̄3̄	2)1̄4̄	5)4̄5̄
6)2̄4̄	3)2̄7̄	6)3̄0̄	2)1̄6̄	3)2̄4̄		

2

A 13)1̄1̄8̄5̄

B 8)4̄0̄,4̄1̄2̄

C 5)6̄7̄8̄

D 62)7̄9̄1̄

E 6)2̄0̄4̄4̄

F 90)1̄4̄0̄0̄

3

A There are 8 apartments in each building. There are 96 apartments. How many buildings are there?

B There are 1269 houses in my hometown. There are 27 houses on every block. How many blocks are in my hometown?

C Janet has 37 carrots. She cooks 20 for dinner. How many carrots does she have left?

D Kelly learned 33 new words every month. She learned 1980 new words. How many months did Kelly learn new words?

E Ray put 7 glasses on each tray. How many glasses did he put on trays if there are 385 trays?

F A farmer had 1060 cows. 7 ran away. How many remain?

4

A 44⟌5990

B 7⟌883

C 24⟌2165

1

8)24   7)35   9)18   5)35   6)36   7)56   8)72

6)42   5)20   4)32   7)28   8)32   9)45   6)18

9)72   7)49   8)16   9)36   5)30   6)24   7)21

6)54   9)27   5)25   4)40   8)48   6)12   9)63

7)42   6)48   8)40   5)40   7)63   6)30   4)16

9)54   4)24   8)56   7)14   8)64

2

A A train made 8 trips every month. It made 888 trips. How many months did the train make trips?

B Yesterday our dog rolled over 14 times. Today it rolled over 35 times. How many times did our dog roll over in all?

C There are 32 people in each building. There are 187 buildings. How many people are there?

D Every rug has 28 pins in it. There are 896 pins in rugs. How many rugs are there?

Part 2 continues on the next page.

E Mindy has 247 grasshoppers. 192 hop
away. How many grasshoppers are left?

```
┌─────────┐
│         │
└─────────┘ ─ ─ ─ ─ ─ ─ ─ ─ ─ ─
```

F Chuck has 1050 puzzle pieces. Every
puzzle is made out of 50 pieces. How
many puzzles does Chuck have?

```
┌─────────┐
│         │
└─────────┘ ─ ─ ─ ─ ─ ─ ─ ─ ─ ─
```

3

A 23$\overline{)2174}$ B 17$\overline{)886}$ C 27$\overline{)1242}$ D 7$\overline{)9827}$ E 23$\overline{)4181}$

F 13$\overline{)328}$ G 4$\overline{)2426}$ H 15$\overline{)7642}$ I 3$\overline{)8859}$

1

$$4\overline{)16} \quad 6\overline{)12} \quad 7\overline{)63} \quad 3\overline{)27} \quad 5\overline{)45} \quad 6\overline{)42} \quad 8\overline{)24}$$

$$6\overline{)54} \quad 4\overline{)12} \quad 9\overline{)54} \quad 3\overline{)9} \quad 7\overline{)56} \quad 10\overline{)20} \quad 6\overline{)36}$$

$$1\overline{)7} \quad 3\overline{)24} \quad 4\overline{)8} \quad 5\overline{)25} \quad 6\overline{)18} \quad 8\overline{)32} \quad 3\overline{)15}$$

$$10\overline{)100} \quad 8\overline{)16} \quad 3\overline{)6} \quad 4\overline{)20} \quad 1\overline{)4} \quad 6\overline{)30} \quad 4\overline{)28}$$

$$3\overline{)21} \quad 5\overline{)10} \quad 4\overline{)24} \quad 6\overline{)24} \quad 3\overline{)12} \quad 7\overline{)49} \quad 6\overline{)48}$$

$$3\overline{)18} \quad 4\overline{)36} \quad 9\overline{)81} \quad 9\overline{)72} \quad 4\overline{)32}$$

2

A $\quad 15\overline{)915}$ B $\quad 37\overline{)889}$ C $\quad 7\overline{)780}$ D $\quad 3\overline{)6924}$

E $\quad 88\overline{)4604}$ F $\quad 24\overline{)2181}$ G $\quad 11\overline{)1385}$ H $\quad 23\overline{)2000}$

3

A Ernest has 30 boxes. 10 are in stack A. The rest are in stack B. How many are in stack B?

B Jan makes stacks of boxes. She puts 31 in stack A, and 31 in stack B, and 31 in stack C. How many boxes are there?

C Bill can stack 59 boxes in an hour. If he stacks 295 boxes, how many hours did he work?

D Martin has 64 boxes in each stack. 12 are in stack A. The rest are in stack B. How many are in stack B?

E There are 3 boxes in each stack. There are 102 stacks. How many boxes are there?

F Katie can stack 45 boxes in an hour. If she stacks boxes for 5 hours, how many boxes can she stack?

G Ana needs 56 boxes. She already has 12. How many boxes does she still need to get?

Transition Lesson 6

1

A
$$2\overline{)6} \quad 3$$

B
$$5\overline{)10} \quad 2$$

C
$$2\overline{)12} \quad 6$$

2

A
$$5\overline{)15}$$

B
$$5\overline{)10}$$

C
$$5\overline{)20}$$

D
$$5\overline{)5}$$

3

A
$$4\overline{)24} \quad 6 \qquad 4\overline{)24} \quad 6 \qquad 6\overline{)24} \quad 4$$

B
$$3\overline{)15} \quad 5$$

C
$$4\overline{)12} \quad 3$$

4

A $\qquad$ B $\qquad$ C $\qquad$ D

$$5\overline{)20} \qquad 2\overline{)6} \qquad 2\overline{)12} \qquad 5\overline{)15}$$

5

A $\qquad$ B $\qquad$ C

Transition Lesson 27

1

A
$$2\overline{)6} \quad \frac{3}{}$$

B
$$5\overline{)10} \quad \frac{2}{}$$

C
$$2\overline{)12} \quad \frac{6}{}$$

2

A
$$4\overline{)24} \quad \frac{6}{} \qquad 4\overline{)24} \quad ^{6} \qquad 6\overline{)24} \quad ^{4}$$

B
$$3\overline{)15} \quad \frac{5}{}$$

C
$$4\overline{)12} \quad \frac{3}{}$$

3

A
$$5\overline{)6\ 3\ 8}$$

B
$$5\overline{)4\ 6\ 2\ 7}$$

C
$$5\overline{)5\ 3\ 2}$$

D
$$9\overline{)4\ 6\ 7}$$

4

A
$$5\overline{)2\ 1}$$

B
$$5\overline{)1\ 2}$$

C
$$5\overline{)1\ 6}$$

D
$$5\overline{)9}$$

E
$$5\overline{)1\ 9}$$

5

A
$$9\overline{)2\ 1}$$

B
$$9\overline{)1\ 2}$$

C
$$9\overline{)3\ 9}$$

D
$$9\overline{)3\ 2}$$

E
$$9\overline{)1\ 1}$$

Mastery Test Review—Lesson 20

1

A	B	C	D	E	F
5⟌1 8	9⟌1 9	5⟌1 7	9⟌6	9⟌3 8	5⟌6

2

A
```
      2
9⟌2 2
 − 1 8
```

B
```
      3
9⟌3 3
 − 2 7
```

C
```
      4
9⟌4 4
 − 3 6
```

3

A	B	C	D	E
9⟌3 8	5⟌1 4	5⟌2 3	9⟌1 7	9⟌2 9

4

A
```
      5
9⟌5 2
 − 4 5
```

B
```
      2
9⟌2 3
 − 1 8
```

C
```
      4
9⟌4 0
 − 3 6
```

Mastery Test Review—Lesson 27

A
A dog chewed 6 bones every day. It chewed bones for 90 days. How many bones did the dog chew?

B
George cleaned 9 erasers each week. He cleaned 45 erasers. How many weeks did George clean erasers?

C
Marcy washed 3 cars every hour. She washed cars for 60 hours. How many cars did Marcy wash?

2

A

Shu fixed 4 tires every day. He fixed tires for 32 days. How many tires did Shu fix?

B

Mr. Silbert wrote 2 stories every week. He wrote 12 stories. How many weeks did he write stories?

C

Carmen painted 5 cars every week. She painted 10 cars. How many weeks did Carmen paint?

Mastery Test Review—Lesson 30

1

A

5)2846

B

9)4730

C

3)7305

D

9)3790

E

9)1231

F

5)2173

2

A   B   C   D   E

9) 2 0 8 1   8) 6 9 3 7   8) 7 9 0 4   5) 2 5 9 6   8) 4 4 7 0

Mastery Test Review—Lesson 35

1

A
```
      4
9) 3 6 2 4
  -3 6
      0
```

B
```
      5
8) 4 0 6 5
  -4 0
      0
```

C
```
5) 4 5 3 0
```

D
```
3) 1 2 1 0
```

E
```
9) 1 8 3 8
```

2

A
```
      4
5) 2 0 3
  -2 0
      0
```

B
```
      5
7) 3 5 6
  -3 5
      0
```

C
```
7) 7 4
```

D
```
2) 1 6 1
```

E
```
2) 8 1
```

F
```
7) 2 1 3
```

Mastery Test Review—Lesson 39

1

A 1 4 6   B 3 9 1   C 6 0 5   D 9 9 2

2

A 798 rounds to _ _ _ _ _ _ tens.   B 474 rounds to _ _ _ _ _ _ tens.

C 203 rounds to _ _ _ _ _ _ tens.   D 645 rounds to _ _ _ _ _ _ tens.

E 155 rounds to _ _ _ _ _ _ tens.   F 297 rounds to _ _ _ _ _ _ tens.

G 579 rounds to _ _ _ _ _ _ tens.   H 512 rounds to _ _ _ _ _ _ tens.

3

A 265 rounds to _ _ _ _ _ _ tens.   B 304 rounds to _ _ _ _ _ _ tens.

C 737 rounds to _ _ _ _ _ _ tens.   D 972 rounds to _ _ _ _ _ _ tens.

E 180 rounds to _ _ _ _ _ _ tens.   F 846 rounds to _ _ _ _ _ _ tens.

G 591 rounds to _ _ _ _ _ _ tens.   H 459 rounds to _ _ _ _ _ _ tens.

Mastery Test Review—Lesson 47

1

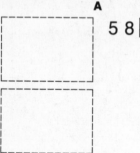

A 5 8 |1 8 7 2

B 7 2 |9 8 4

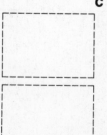

C 4 8 |9 4 8

D 6 2 |2 5 6 1

Mastery Test Review Lesson 47 (continued)

2

A $57\overline{)752}$

B $22\overline{)946}$

C $39\overline{)2528}$

D $70\overline{)1726}$

Mastery Test Review—Lesson 50

1

A $22\overline{)794}$

B $35\overline{)3243}$

C $56\overline{)2327}$

D $51\overline{)2481}$

E $72\overline{)3829}$

2

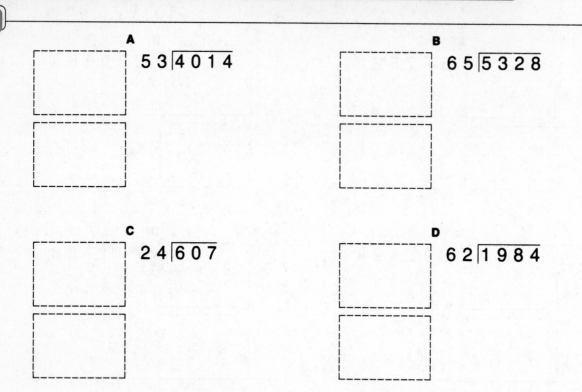

A

$53\overline{)4014}$

B

$65\overline{)5328}$

C

$24\overline{)607}$

D

$62\overline{)1984}$

Mastery Test Review—Lesson 54

1

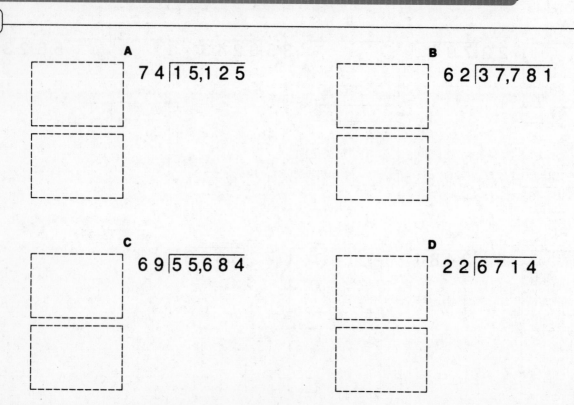

A

$74\overline{)15,125}$

B

$62\overline{)37,781}$

C

$69\overline{)55,684}$

D

$22\overline{)6714}$

2

A

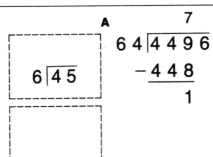

$$\begin{array}{r} 7 \\ 6\,4\overline{\smash{)}4\,4\,9\,6} \\ -4\,4\,8 \\ \hline 1 \end{array}$$

B

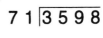

$$7\,1\overline{\smash{)}3\,5\,9\,8}$$

C

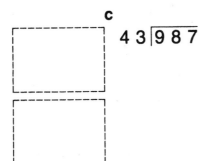

$$4\,3\overline{\smash{)}9\,8\,7}$$

D

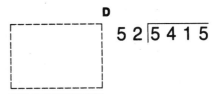

$$5\,2\overline{\smash{)}5\,4\,1\,5}$$

Mastery Test Review—Lesson 60

1

A There are 6 kids on each team. There are 48 kids. How many teams are there?

       +   −

       ×   ÷

B Sherry has some leaves. 8 of them are red and 9 of them are yellow. How many leaves does Sherry have?

       +   −

       ×   ÷

C The parade has 17 bands. 6 of the bands are on floats and the rest of them are marching. How many bands are marching?

       +   −

       ×   ÷

Part 1 continues on the next page.

D Pete did 35 problems on each page. He did 11 pages of problems. How many problems did he do?

```
┌ ─ ─ ─ ┐                          +    −
│       │ ─ ─ ─ ─ ─ ─ ─ ─ ─ ─ ─ ─ ─
└ ─ ─ ─ ┘                          ×    ÷
```

E Betty typed 7 pages an hour. She typed 56 pages. How many hours did she type?

```
┌ ─ ─ ─ ┐                          +    −
│       │ ─ ─ ─ ─ ─ ─ ─ ─ ─ ─ ─ ─ ─
└ ─ ─ ─ ┘                          ×    ÷
```

2

A A woman has 126 plants in her garden. Her daughter has 96. How many more plants does the woman have than her daughter?

```
┌ ─ ─ ─ ┐                          +    −
│       │ ─ ─ ─ ─ ─ ─ ─ ─ ─ ─ ─ ─ ─
└ ─ ─ ─ ┘                          ×    ÷
```

B Vic drinks water 6 times every day. After 42 days go by, how many times will Vic have drunk water?

```
┌ ─ ─ ─ ┐                          +    −
│       │ ─ ─ ─ ─ ─ ─ ─ ─ ─ ─ ─ ─ ─
└ ─ ─ ─ ┘                          ×    ÷
```

C There are 4 cans in each box. There are 24 cans in boxes. How many boxes are there?

```
┌ ─ ─ ─ ┐                          +    −
│       │ ─ ─ ─ ─ ─ ─ ─ ─ ─ ─ ─ ─ ─
└ ─ ─ ─ ┘                          ×    ÷
```

D The doctor saw 32 patients on Monday, 20 patients on Tuesday, 4 patients on Wednesday, and 7 patients on Thursday. How many patients did she see?

```
┌ ─ ─ ─ ┐                          +    −
│       │ ─ ─ ─ ─ ─ ─ ─ ─ ─ ─ ─ ─ ─
└ ─ ─ ─ ┘                          ×    ÷
```

Part 2 continues on the next page.

E Every bag Mike carried weighed 18 kilograms. If he carried 9 bags, how many kilograms did he carry?

```
┌ ─ ─ ─ ┐
│       │          _ _ _ _ _ _ _ _ _ _ _ _ _ _ _ _ _
└ ─ ─ ─ ┘
```

                                                    +   −

                                                    ×   ÷

F Millie went to the bank 9 times every month. In all, she went to the bank 126 times. How many months did Millie go to the bank?

```
┌ ─ ─ ─ ┐
│       │          _ _ _ _ _ _ _ _ _ _ _ _ _ _ _ _ _
└ ─ ─ ─ ┘
```

                                                    +   −

                                                    ×   ÷